T0339533

Mastering Project Discovery

Introducing a comprehensive approach to invigorate project leadership, this book provides a framework – the OUtCoMES Cycle – for developing, managing, advancing, and optimizing engineering and analytics projects.

All too often, issues of moral hazard and completion bias prevent engineering and analytics managers and team leaders from asking the critical question, 'What's the problem?', before committing time, energy, and resources to solve it. This book draws attention to the definition, structuring, option consideration and ultimately the addressing of the right problems, exploring the OUtCoMES Cycle framework that facilitates and energizes systematic thinking, knowledge sharing, and on-the-fly adjustment with an explicit focus on the maximization of value and ROI. Each chapter includes discussions and lessons in analytical and engineering problem identification, problem structuring, iterative problem development (mental and computational) and problem resolution, at least three embedded real-world case studies, and a closing 'Practitioner's Recap' to contextualize key chapter takeaways.

Written by a team of established academic scholars and practicing analysts and engineers, this is an accessible and culture-shifting action guide for instructors interested in training the next generation of project and analytics leaders, students of analytics and engineering, as well as practicing project leaders and principals.

Elliot Bendoly is the Fisher College of Business Distinguished Professor at The Ohio State University and has served as Editor-in-Chief of the *Journal of Operations Management*. Apart from more than 70 peer review articles, he has authored eight texts including *Excel Basics to Blackbelt* and *Visual Analytics for Management*.

Daniel G. Bachrach is an elected Fellow of the American Psychological Association (APA), the Association for Psychological Science (APS), and the Society for Industrial and Organizational Psychology (SIOP). He is a

Business Professor at the University of Alabama and his research focuses on the structure and movement of information in teams.

Kathy Koontz is a Principal Data Strategist with Amazon Web Services, working to help customers create sustainable competitive advantage from their data, analytics, artificial intelligence, and machine learning investments. She has more than 30 years of experience working with large organizations across multiple industries in both business and technology roles.

Porter Schermerhorn is a Principal Engineer at Amazon where he holds patents across AWS, marketing, and customer engagement. A career mentor and coach, Porter serves in the MBA Mentorship Program at the University of Washington and on the advisory board for the Russ College of Electrical Engineering and Computer Science at Ohio University.

Mastering Project Discovery

Successful Discipline in Engineering and
Analytics Projects

**Elliot Bendoly, Daniel G. Bachrach,
Kathy Koontz, and
Porter Schermerhorn**

NEW YORK AND LONDON

Designed cover image: The cover image was provided by the professional photography services of Laura Bendoly, for whom we are greatly appreciative. Her work captures the synergies between clear structure and potential alternatives, so fundamental to the discovery process that the OUtCoMES Cycle is designed to facilitate.

First published 2024
by Routledge
605 Third Avenue, New York, NY 10158

and by Routledge
4 Park Square, Milton Park, Abingdon, Oxon, OX14 4RN

Routledge is an imprint of the Taylor & Francis Group, an informa business

Library of Congress Cataloguing-in-Publication Data
Names: Bendoly, Elliot, author.
Title: Mastering project discovery : successful discipline in engineering and analytics projects / Elliot Bendoly, Daniel G Bachrach, Kathy Koontz, and Porter Schermerhorn.
Description: New York, NY : Routledge, 2024. | Includes bibliographical references and index. |
Identifiers: LCCN 2023049463 (print) | LCCN 2023049464 (ebook) | ISBN 9781032548210 (hardback) | ISBN 9781032548197 (paperback) | ISBN 9781003427650 (ebook)
Subjects: LCSH: Software engineering--Management. | Information technology projects--Management. | Project management.
Classification: LCC QA76.758 .B4624 2024 (print) | LCC QA76.758 (ebook) | DDC 005.1068--dc23/eng/20231214
LC record available at https://lccn.loc.gov/2023049463
LC ebook record available at https://lccn.loc.gov/2023049464

ISBN: 978-1-032-54821-0 (hbk)
ISBN: 978-1-032-54819-7 (pbk)
ISBN: 978-1-003-42765-0 (ebk)

DOI: 10.4324/9781003427650

Typeset in Times New Roman
by MPS Limited, Dehradun

Contents

Foreword

Thomas H. Davenport

Many things are changing in the worlds of systems engineering, analytics, and AI. With generative AI, for example, we now have the ability to create extensive programming or data analysis with only a short prompt. Nontechnical users can also employ "low code/no code" tools to create departmental-level applications with only a few clicks. Automated machine learning systems can evaluate multiple algorithms within seconds. What has not changed, however, is the importance of understanding the problem that needs to be solved. The ultimate objective of the program or analysis that one is trying to create is not something that an AI system can determine; that requires context, consultation, communications, and clear thinking by humans. These topics are the focus of this book. The OUtCoMES cycle it describes is key to solving the right problem for the organization. In an era when it is becoming much easier to solve data-oriented problems, it is even more important to ensure that engineers, analysts, and data scientists are working on the issues and solutions that matter most.

Thomas H. Davenport, Distinguished Professor,
Babson College, Fellow, MIT Initiative on the
Digital Economy, Senior Advisor, Deloitte AI Practice

Foreword

Jack Phillips

If you're reading this book, then most likely you've spent your fair share of time in and around analytics and engineering projects, either as a member of a project team or as a team leader, or more than likely both at various points in your career. One of the things that you've probably seen too often, as I have, are lost opportunities to generate a conclusive win. And you've probably seen these even when wins seem like they are well within reach. Why is that? Success in engineering and analytics projects often boils down to some really not-immediately-obvious elements that tend to lurk around the edges of every project's day-to-day focus. Is the team working on the most important problem? How can you generate evidence and confidence that the team is working on the most important problem, from a set of given alternatives? Can the project teams that follow in the wake of this current project find specific ways to accelerate their own success by building directly or indirectly from the prior team's work? These are important questions that all analytics and engineering projects have to find answers to in order to build and to maintain a culture of project team success. As the CEO and Co-founder of the International Institute for Analytics (IAA), I regularly see first-hand the tremendous value that's created when organizations have a culture that emphasizes and systematically focuses on process and execution excellence in analytics and engineering projects. On the flip side, I also see the tremendous waste of effort, and the enormous opportunity costs when they don't approach project execution with this kind of systematic, focused discipline.

One of the topics that has tended to come up the most frequently in and around the analytics and engineering projects that I've had both direct and indirect exposure to throughout my career is the question of identifying the most effective applied frameworks for consistently driving project excellence, project after project. Now for the first time in Mastering Project Discovery (MPD), this book introduces a critical, encompassing framework, what's called the OUtCoMES Cycle (The Cycle), that is specifically designed to minimize wasted effort and the risk of project failure. The "Cycle" provides project leaders and members tasked with project co-leadership with a straightforward, easy to implement, structured project discovery and development process that is tied directly to objectives that maximize project value and practical impact. The sum total of project value across an enterprise, of course, is the key to industry differentiation and generating sustainable competitive advantage. With the Cycle, MPD offers a timely, practical, and easy-to-implement approach for project leaders tasked with solving the right problem in the right way, and generating value from analytics and engineering projects. Importantly, the Cycle also creates an organizational record of process considerations and decisions, providing a systematic guide and source of problem insight for use by subsequent project teams operating in the same or adjacent spaces.

In developing the Cycle, the author team draws from decades of practical experience in the field of analytics and engineering in both academia and in industry. The OUtCoMES Cycle was developed through systematic academic study of best practices in project management, combined with practical application across a wide range of different industry settings. In addition, the cases, exercises, and practitioner summaries provided in each chapter offer clear examples of how and where The Cycle can help to predictably reduce risk and to create value. After reading Mastering Project Discovery, I am excited to see how organizations apply these principles to improve their project outcomes. Whether you're a student, project team member or leader, or executive responsible for continuous innovation and value creation, I have every confidence that you'll find value in applying the approach described in The Cycle to your organization's analytics and engineering projects.

Jack Phillips, CEO and Co-Founder of
International Institute for Analytics (IIA)

Author's Overview

Getting the most out of an engineering or analytics project requires the thoughtful coordination of specifically targeted solution tactics. However, generating optimal returns from this process also critically presumes that the right problem has been targeted. How do we increase the chances of getting that right, and of advancing on practical solutions to address the right problems? How do we do this while also allowing ourselves the opportunity to step back, and (perhaps fundamentally) redefine the problem, when we reach diminishing or even negative returns on our investment of time and resources? The primary aim of this book is to provide an integrated, comprehensive framework and structured process to help manage, advance, and optimize engineering and analytics projects – what we call the OUtCoMES Cycle. We motivate and provide practical anchorage of the Cycle using a series of exercises and real-world case examples, highlighting the importance of key principles and organizational architecture. For color, we include related historical and literary references, and provide closure to each chapter in the form of a 'practitioner's recap'. The collaboration between established academic scholars, analysts, and engineers at Amazon makes for a ground-breaking and culture-shifting practical reference and guide for students, practitioners, and academic scholars alike. The book both offers directions for building deep understanding of how to repeatedly identify and address the right problem within project contexts, as well as practical resources to help readers master this process.

Introduction

"What's the Problem?"

What's wrong with the way managers, engineers, and analysts work together to find solutions?

If you're currently occupying one these roles, and reading this question, you'll probably have a hard time choking back a laugh. You might even be tempted to ask the question, "What do you mean 'work together'?" Or, even more cynically, "What solutions?". Going a step further, if you've ever spent time thinking about these questions, you may have also started to develop some suspicions about where the blame for NOT working together, or NOT finding solutions, probably lies. Maybe the wrong people were involved in the project in the first place? Maybe the incentives for the project were misaligned, or introduced conflicts with other key stakeholders? It can be very hard to pinpoint exactly what accounts for roadblocks and bottlenecks that keep teams from finding success. And, it's only natural to try to figure out why projects don't deliver. Doing this, autopsying failures, is an important underlying element of the sensemaking process. It allows us to reconcile conflicting realities that separate our own sense of what 'should' have happened with the project versus the reality of what actually 'did' happen. 'Should' versus 'did' is a theme we'll return to repeatedly.

Unfortunately, blaming-others – what happens in most organizations – rather than digging more deeply into critical process issues, is not likely to bear fruit in the long run. Even the smartest, incentive-aligned managers, engineers and analysts can find themselves wondering about project ROI – return on investment. These professionals could as easily point their fingers at one another to explain why the team failed as at some invented cause. But, this finger-pointing is unlikely to solve anything in the long-term because, more than likely, these same individuals will be on similar – or even the same! – teams down the road. Members will be operating with the same underlying assumptions, expectations and biases and generating the same kinds of disappointing results. Experts with complimentary domains of expertise, who have to coordinate their efforts to accomplish multi-faceted objectives,

DOI: 10.4324/9781003427650-1

depend on one another in fundamental ways. As a core matter of process, some goals simply can't be achieved without specialized experts, with diverse sets of knowledge and skills, pulling together.

An important question that we seek to answer, then, is why don't organizations investing in teams of well-paid, incentive-aligned professionals always get outstanding practical solutions from their smart, well-trained, well-intentioned, motivated experts who are all working toward the same goal? Optimal solutions – or even functional/acceptable solutions – are often elusive because of the natural challenges associated with transparent communication. While these smart, well-trained, well-intentioned and motivated professionals may be very good at their management, engineering and analytics jobs, they also may have a great deal of difficultly communicating: (1) their needs or (2) their capabilities across functional fault lines. Among other adverse outcomes, this kind of communication breakdown can result in what is referred to in academic parlance as a moral hazard. A moral hazard is a situation where individuals feel insulated from the consequences of their actions. Engineers and analysts who have marching orders from management, and who are motivated to work towards what they believe they were tasked with achieving, are disinclined to think beyond the assumptions and mental models tied to their own specific function. Why would they?

Yet, because operational transparency is more realistically a smudgy opacity at best, results from concerted efforts may or may not have much cross-functional value. For the same reason, because communication and coordination are sluggish at best and frozen at worst, it can often be very difficult to pinpoint just what went wrong, when it went wrong, and how to fix it. This procedural and operational haziness protects engineers and analysts against consequences – the moral hazard problem – which in itself has both pros and cons. It can, of course, promote the creativity and risk taking that is the lifeblood of radical innovation because it frees experts to follow leaps of intuition without the constraint of extraneous stakeholder goals. But, it can also promote waste, neglect, carelessness, and insensitivity. It can promote ill-used time, energy, effort, and resources that ultimately fail to deliver a consistently high rate of return. This declining cycle can in turn lead to satisficing maneuvers that leave all vested stakeholder interests with unsatisfied goals and unmet expectations. This sequence of diminishing returns is, of course, unfortunately very familiar to many of you reading this book. And, it yields an all too familiar story – salvage efforts to make lemonade out of the lemons that, though neither requested nor targeted for harvest, were nevertheless delivered by the team.

Beyond these filtered and obscured cross-functional communication and coordination challenges, however, something more fundamental is often at work as well. And, it will also be no surprise.

The fact is that, for the most part – and across a wide range of project scenarios similar to those that we describe as case examples within the chapters of this book – management, engineering and analytics team leaders seldom give themselves enough time to ask the critical question 'what's the problem?', before committing time, energy and resources to solve it. Ready, shoot, aim ... repeat. There is an often-made assumption, during the team assembly phase of the project development process, that can fundamentally short-circuit necessary critical reflection. The assumption is that the deep and broad experience of top-level managers and project leaders enables them to, with near-perfect accuracy, determine the nature of the problems that are worth solving, or that would be important to solve, and that identifying solutions is simply a matter of follow-through on the part of the project team tasked with solving it. Completion bias, or a tendency to seek out the satisfaction of closure, for example, regarding organizational improvement discussions, places solution-finding above important deliberations in problem development. In other words, finding a solution to *some* problem becomes more important than being sure that the right problem was identified in the first place. In light of this reality, the grim statistics that Goldman and Taylor (2023) report bearing on project failures (e.g., 64% meeting stated goals) should come as no surprise.

In an analogy that we like to reference in front of student and practitioner audiences alike, however, problem solving isn't simply finding your way through a maze. It also, critically, involves developing an understanding of the structure of the maze itself. Like a maze carved into so many jigsaw puzzle pieces, and tossed up into the air – connecting the pieces, getting to that understanding of structure, can be a formidable challenge. This is especially true when, as is often the case, it turns out that many of the pieces are missing. But, imagine trying to solve the maze prior to that assembly. What are the odds of starting in the wrong place? Ending in the wrong place? Missing critical barriers or shortcuts? In contrast, working your way through and ultimately 'solving' the maze once a model of that structure – or even a sufficient partial model of the structure – has been pieced together can prove far, far easier. Having a functional map makes getting to the right destination much easier.

A related assumption that is often made is that there is an essentially linear path between the status quo – i.e., the current state of the ecosystem in which the problem is embedded – and the understood value-added exercise the team is tasked with executing. Or, at least, that covering the distance between point A and point B adds value in a relatively monotonic way. Challenges that come up along the path moving from point A to point B are frequently viewed as challenges to overcome, with brute force if necessary. Those initiating project work in the first place, in a worst-case scenario, may even have an *a priori* solution in mind, such that the project serves as little more than a confirmatory validation exercise – and not necessarily with the full consent and knowledge of the team executing the project.

Unfortunately, this worst case turns out to be not all that uncommon. Not surprisingly, it also tends to be extremely costly financially, emotionally, and culturally for both the team executing the project, as well as for the organizational ecosystem in which the team is embedded, impacting numerous stakeholders along the way. Even good faith efforts undergirded by a discovery motive can fail as a result of misdirection at any stage in the process leading up to problem determination. Picking the wrong path to start out on, as a point of departure, represents an expensive misstep early-on in this process. But, then also refusing – or being unable – to adjust course, change tack, back-track, or even start over when it becomes evident the initial path is fruitless, also is a fail. Organizations waste too much time coming up with impractical solutions in large part because poor problems – or the wrong problems – have been identified, privileged, and committed to by resource gatekeepers and policy makers.

What can well-intentioned professionals do to help solve this engrained cultural myopathy? How can we avoid or rectify a failure to understand one another or the context in which the team is embedded? What can we do to guard against getting locked into focusing on the wrong problems, failing to identify issues driven by communication breakdowns and functional boundary hurdles, missing out on opportunities to maximize returns from the tremendous investments made regularly in the wrong projects?

The key to dealing with this common, expensive, and culturally engrained set of related issues, as it almost always is when confronting serially complex, unstructured problem spaces … is structure – structure with a capital "S" rather than a small "s". Functional structure rather than inhibiting structure. Structure that channels rather than structure that constrains. Structure that can be used as a guide to help integrate systematic thinking into a team's processes, rather than excessively constraining the team's inquiry. Structure that highlights alternatives, and facilitates license to explore, fail, learn, reassess and adjust mid- or even late-course onto a more productive track. Structure that provides a common platform for communicating the best ideas at any given stage in a process, but that also tracks the intersection, interpretation, application and development of those ideas for retrospective consideration and re-evaluation. Structure that facilitates transcendence of functional boundaries, and encourages the integration and coordination of expertise across established functional and experiential fault lines, inspiring a culture of discovery.

That is what this book is about.

In the chapters that follow, we examine problem development and solution tactics. We examine where the development of problems can fall short, how to avoid such missteps and how to get the most out of the time and resources invested in problem-solving projects. Central to this mission, we present and discuss a framework and process for structuring problems, outlining alternative problem specification and solutions, and getting to practical and high-return solutions: The OUtCoMES Cycle. The approach facilitates and energizes systematic thinking, knowledge sharing, and on-the-fly adjustment with an explicit focus on the maximization of value and ROI. The latter chapters delve into more explicit tactics for leading effective analytical and engineering projects, fostering systems of transactive memory that can generate ongoing dividends for teams tasked with leveraging diverse knowledge and expertise to solve novel and complex problems.

Part I

Virtuosity

Chapter 1

The Value of Structure and Alternatives

Our motivation to help analysts, engineers, and managers discover the best projects is borne out of the most fundamental of challenges. It's a challenge that we see industry practitioners facing time and time again: namely, amid the chaos that is day-to-day operations, not knowing where or how to start something new in the absence of clear structure. We can get a sense for the pervasiveness of this challenge through quotes drawn from popular culture. For example, consider the following statements:

> *Out of clutter, find simplicity. From discord, find harmony. In the middle of difficulty lies opportunity.*
> *— Albert Einstein/John A. Wheeler*[1]

> *In limits, there is freedom. Creativity thrives within structure.*
> *— Julia Cameron, Mark Bryan, The Artist's Way, 1992*

> *I thrive in structure. I drown in chaos.*
> *— Anna Kendrick, Scrappy Little Nobody, 2016*

The increasingly complex challenges employees are regularly tasked with addressing at work, just like those that we all face outside of work in our personal lives relating to finances and family and social relationships and goals and plans for the future, emerge from interactions and assumptions and interdependencies and limitations situated across a wide range of factors embedded in a complex and evolving reality. Two key hurdles to successfully overcoming both these work and life challenges is first understanding, and then effectively integrating, a large number of interacting parts. Many of these parts are beyond our control. Some of these parts are difficult to define. Some of these parts become salient to understanding or defining what the broader picture looks like at different

DOI: 10.4324/9781003427650-3

times – or for different lengths of time. It is often unclear how one part of this broader ecosystem impacts the relevance or performance of other parts of the system. It can be unclear at the outset of deliberations how to weigh or value these emerging and interacting parts, and even if these weights were 'known', it can be difficult to determine what the most critical, or set of most critical, performance measures is likely to be once a decision is made.

This complexity, dynamism, and potential equifinality applies as much to the challenge of strengthening the long-term environmental sustainability a transnational organization's supply chain as it does to successfully fostering healthy life goals and long-term well-being of one's own children. In both sets of complex systems, in which the structure of the relationships between interdependent and dynamic parts are continually evolving and changing, there are camouflaging temporal lags between platonic cause and effect. There is no clear answer to the question of when we should expect actions to yield measurable or relevant impact. False signals obscure logical and effortful assessments of cause-and-effect. Can we ever really know what we did that made the difference in efforts to orchestrate a raw materials purchase at just the right arbitrage moment, or for that child who was on the cusp of quitting the soccer altogether because she didn't make the varsity team, but decided to stick it out at the very last minute? There can even be competing – virtuous – priorities at a given moment in a process that lead us to question whether a seemingly smart move or set of decisions, effectively advancing an agenda towards a prioritized outcome (soccer or the robotics team …!?), might be the best use of available resources. How do well-intentioned stakeholders, with the best interests of all of the key (known) players squarely in focus, contend with this complexity, and ultimately end up with a solution that is both functional and realistic?

The short answer: we simplify. This is just something that human beings are exceptionally well-equipped to do. Let's do a little in-the-moment analysis of this very human process to let this assertion – that we simplify – play out in real time. Just take a second, right now, to count the number of sounds in the room where you're sitting that you are unconsciously ignoring at this very moment – the fan, the breathing of your spouse or pet – or your own breathing, the whir of the equipment in your office, the creaking of your chair, the traffic sounds outside on the street, the birds in the tree outside your office, the sound of the airplane's engine outside the window, … and on and on and one. We are very good at simplifying the inflow of signals bombarding us almost continuously – otherwise, we'd simply be overwhelmed with (for the most part) otherwise distracting noise and data that would keep us from focusing in on the important parts of the system.

Either consciously or unconsciously, we are highly adept at filtering out the vast majority of the reality in which our everyday challenges are situated. What we also do very well is generalize relationships and effects. Traffic sounds and birds, we unconsciously place in a virtual box labeled 'outside'. Weighting and considering options for strengthening the sustainability of a transnational's supply chain, within a network of suppliers situated in locations with green-regulations defined by varying degrees of potency and relevance, we group and categorize certain kinds of emissions into bins with different levels of decision relevance. We weigh and compare the complex parts that define our operating spaces in terms of months and years, not in terms of minutes and seconds. We implicitly – automatically – assume that actions and steps and policies that we've been able to tie (perhaps justifiably, perhaps unjustifiably ...!) to outcomes in the past can and will yield similar outcomes in the future. Most importantly, we form efficient rules of thumb – in academic parlance, we refer to these simplifications as heuristics – when we need to come up with quick problem formulations and solutions, to repeatedly encountered scenarios. And, with these satisficing efforts based on imperfect and incomplete information, we put these necessarily expedient measures in place, cross our fingers and we hope for the best, with no way of knowing in advance whether we should have chosen option A instead of option B. Should I have encouraged her to put in one more season on the JV soccer team, or should I have just encouraged her to jump more heavily into robotics. It can be hard – even impossible – to tell.

And, from an evolutionary perspective – which in aggregate terms generates benefits at the level of group but not necessarily for individuals in all instances ...! – this isn't at all a fundamentally bad tactic. In a broadly generalized survival of the fittest sense, humankind's ability to filter and simplify and act expediently, based on incomplete information, has served us pretty well over the last hundred thousand years or so. Today, the rules of thumb that define many of our programmed decisions – heuristics – generate recognized process efficiencies with clear bottom-line benefits. We transmit these rule-of-thumb heuristics – these simplifications – along to others (both explicitly through codification as well as implicitly through practice and custom). We make our own personal modifications to these simplifying heuristics and, more often than not in both our personal and professional lives get criticized for taking these analytically imperfect (yet immensely efficient)! short-cuts. But, the fact is that these heuristics, despite having the potential to be somewhat unreliable in instances that depart from baseline status quo assumptions, and also despite often yielding suboptimal results (given best-case hopes and expectations), can in fact be exceptionally helpful as we undertake efforts to generate added value. Heuristic decision making can, and often does, provide a great starting point for defining a problem space and setting down a basic blueprint for the structure of the problem and its ultimate resolution. This kind of rule-of-thumb decision also can reduce the risk of becoming mired in inaction by the shear complexity of the range of potential problem spaces that could emerge from a given set of realities. Ready, set, start ...

1.1 The Best Kinds of Problems to Have

When we discuss problem solving with masters-level (e.g., MBA, MS) and undergraduate students, in on-site executive education programs, or on-the-ground with managers, we often adopt the analogy of 'building a better bike' (aka the 'BBB'). Alluring alliteration aside ... although not necessarily avid cyclists most of these stakeholders are very familiar with bicycles, and also have some fairly consistent beliefs about the physiological (i.e., riding a bike is good for you ...!) and environmental (i.e., riding a bike is good for the planet ...!) benefits of bikes. But, most of these stakeholders really haven't spent a lot of time thinking about how to improve on their design. One of the first questions we typically address in the BBB discussion is what would be better in a "better" bike?'. That is, what does "better" actually mean in this context? (Is it a faster bike? Cheaper? Lighter?) And, along a related track, how would you actually go about measuring your improvements to see if the led to a ... Better Bike?

These questions are of course central to any effort to actually improve on the "bike" in a practical sense. But in order to get there, we also need to pull on some of the threads that emerge, and also to remain open to seeing where they lead. For example, if the improvements prioritized by the group are broadly associated with the speed of the bike, we have to ask what is it that makes a bike fast in the first place? Certainly, this kind of performance question orbits the mechanical design of the bike itself. But, addressing the question effectively also depends on an understanding of the characteristics of the bike's riders. Because individual differences impact the functionality of the bike's mechanical design these two parts of the system – the design of the bike and the attributes of the bike's rider – interact and have to be thought about in junction with one another.

Likewise, if the 'better' identified by the group involves enhancing the comfort of the rider, interactions of this type between these parts of the system (i.e., between the attributes of the rider and attributes of the mechanical design of the bike) are also important in addressing the 'better' question as well. A functional understanding of what aspects of the mechanical design and operational configuration of the bike are likely to impact the rider's comfort. Understanding of whether and to what extent these parts of the system are likely to contribute positively or negatively to the rider's comfort can only be determined through an understanding of the complex interaction between these parts. So, any reference system for tackling these kinds of questions, relating to the speed of the bike for example, or to the comfort of the rider, must by necessity include consideration of (1) the attributes of the bicycle itself, (2) the individual characteristics of riders of the bike, and (3) the nature of the potential (and potentially complex) interactions between these parts of the system.

Does our ability to address issues relating to the speed or comfort of a bicycle require that we, for example, also have a professional understanding of how races like the Pelotonia or the Tour de France operate? After a certain point, and for a very specific kind of bicycle rider, this kind of professional insight could definitely be useful in advancing questions relating to improving bike speed and/or comfort. What about from an operational or functional perspective? Do we need to understand, for example, how roads or bike-paths are constructed? Even at the extremes of fidelity, this kind of operational/functional knowledge probably wouldn't help to generate much movement on improving either the speed or the comfort of the bike. At a more micro-scale, architecturally, is it important for us to understand the materials science underlying the physical mechanics of the bicycle's components and sub-components? Knowledge of the mechanical properties of candidate materials is probably more than enough to advance both speed and comfort goals. It is unlikely that we'd need to spend too much time in consideration of the manufacturing process by which the wheels' spoke wires were drawn out to generate returns from a discussion focused on improving bike speed and comfort. But, we could always go down these kinds of technically acute paths later on in the process if it became obvious that there was some functional benefit in doing so.

A key take-away that emerges from this exercise is that, whether among the MBAs, Executive Education participants, undergraduate students or managers on-the-ground is that, in any effort to effectively tackle what initially emerges as a design challenge, simplification is critical. If we don't very quickly put up relatively high guardrails that limit both the problems' scope and depth of complexity, there is no realistic chance of getting to an operative solution within a realistic time frame. And, this isn't mere speculation or basic pessimism. It is a fact; and it is ok. Without this kind of systematic reduction in system complexity, our ability to make any kind of complex decision – or for that matter even routine or relatively simple decisions – can become frozen by a surfeit of inter-tangled figure-ground data.

In contrast, a well-contained, context-relevant, and goal-appropriate reference system has the potential to get us close to a truly remarkable solution, and very quickly. We just need to accept the fact that making simplifications, using heuristics in problem and solution development, doesn't necessarily guarantee the best problem definition or downstream solution. At least not immediately. Problem simplification and heuristics by definition are limited by system omissions. Missing parts, and missing relationships. Sometimes, we start in the wrong place. Other times, our systematic reductions of reality lead us down the wrong path. It is the capacity to leverage simplification and heuristics, but also to recognize their inherent short-comings and seek alternative vantage points and approaches, that distinguishes effective and ineffective project execution. It distinguishes the best problems, and by extension the best projects, those that return the most value from the resources devoted to them, and that also have an outstanding problem structure at their core. We refer to these kinds of projects as 'exemplar project', because, if not for proprietary restrictions, these are exactly the kinds of projects that organizations would love to see repeated over and over again.

So, how do we maximize our chances of getting this process right? How can we make the most of simplified reference systems? How do we embrace the use of heuristics, but also simultaneously embrace the internal referencing and scrutiny that these bounded approaches demand? How do we maintain a clear roadmap for reconsideration, anchored by authentic and realistic state-of-the problem depictions? Ultimately, the best approach capitalizes on two fundamental elements: *structure* and *alternatives*.

In general, when individuals select specific problems to resolve, either directly or through delegation to others, that choice is drawn from a host of 'alternatives'. Other problems could have been chosen – and likely were chosen by other stakeholders and other times. It also implies and is often, in fact, motivated by specific assumptions regarding problem 'structure'. These two terms, 'structure' and 'alternatives', are reflective of key themes in this book. *Structure* specifies measurable outcomes and the means to advance them. It is how we build problems, and ultimately what allows us to solve them. *Alternatives* refers to the many options we face in the specifics of that structure. This includes what measure or measures tells us know that we've solved a problem, or how well we've solved it?, what levers can help to drive that solution, and how do these levers impact solutions?

Your immediate impression might be that *structure* and *alternatives* are likely to be odds with one another. After all, how can you think outside of the box if you're also expected to keep an eye on the box's corners, walls, and lid at the same time? If you were being asked you to accept the box as-is, without leave to change its dimensions and configuration, or leave it entirely, this could certainly be a problem. But the 'box' we're talking about, as an analogy when we discuss *structure*, is a special box. It's a box built on the premise that *alternatives* (i.e., different ways of thinking about the design of the box, or future boxes) are of value. Working with *structure*, in the way that we outline it in this book, explicitly pushes analysts and engineers to question how that *structure* is used. The walls of the box (i.e., the way we think about the problem) can be pushed out, and the lid lowered as needs arise. There are doors all sides, so that you can move, when compelled by evidence, say, to an adjacent and ostensibly very different box. You can also always return through those same doors. *Structure*, in the way we frame it here, is by no means an impediment to the development of alternatives. Rather, it is an explicit invitation, indeed a means, to encourage the development, close examination, and access to *alternatives*.

To put a sharper point on this, *structure* is almost always – really always – imperfect from the outset. It is imperfect because, in a definitional sense, simplifying complex systems into a manageable subset of parts is difficult. It is the willingness of project leaders, analysts and engineers to update that structure, given a clear understanding of emerging and potential alternatives, that ultimately makes discovery of outstanding solutions to exemplar projects possible. Structure and alternatives provide stakeholders with an adaptive license to fail fast, and often, while systematically advancing exemplar projects. In short, structure and alternatives engender the mastery of discovery.

🚲 *Exercise:* A Little Structure Goes a Long Way

The problem: The role of structure can often go unappreciated. It's impact on how we define problems and subsequently solve them, however, can be very powerful. Take the following example. It an example that most readers of this book – and most people who haven't gotten to it yet …! – are likely to be very familiar with. Packing for a trip. In Figure 1.1., a set of 4 items is displayed as outlines of their relative shape and size. Imagine that you are facing the task of determining how many of these items to place in a single carry-on bag, and which items to place into either the luggage check, or to send as parcels via the mail. Both of the latter incur additional costs per item, and perhaps some additional risk of getting lost in transit. Let's say that your carry-on luggage has the area depicted in Figure 1.2 (a one-third reduction of the rectangular area in Figure 3.1). For simplicity, we assume that no items can overlap, or can be stacked, on one another.

If you weren't concerned about weight, and wanted to get as many of these items into your carry-on bag as you could, what items would you place in that carry-on?

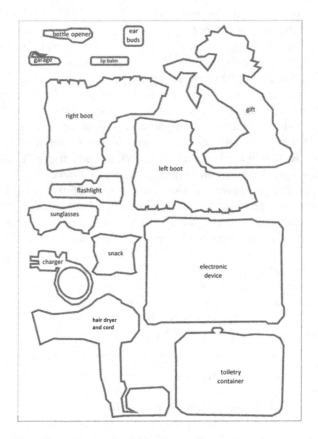

Figure 1.1 An Array of Items Ready for Re-organization.

For this exercise, you might try to cut out the shapes and position them in Figure 1.2. An electronic version of this exercise is available (www.masterdiscovery.com) if you'd like to avoid cutting directly out of this book – but definitely feel free to take any approach you like ☺.

A packing autopsy: If you're satisfied with your arrangement of items at this point, let's do some retrospective breakdown of how you approached the solution you arrived at. What did you move first when packing the reduced space? Typically, participants in the exercise will answer "the largest item", or "the most valuable item". In this case, you might have chosen "the item whose cost or risk of not having in the carry-on was greatest". Some might even say "the item with squarish corners, since I want make sure corners get filled". Any of these answers, and certainly others, would describe

Figure 1.2 A Reduced Space: Which Items Would Fit without Overlap?

one of several very simple rules you've likely used in this process in your own life. They are guiding, rudimentary principles. They aren't necessarily perfect, but they helped you to get started with the process of getting packed.

What did you do for the next few items? Did you stick to the same choice criteria (e.g., try to pack the next biggest thing?), or did you switch to an alternative rule? If so, what rule, and why? Did you rearrange already-packed items, to try new configurations in the space, prior to placing other items into the bag? Both of these second-phase approaches are typical in this exercise. Once again, while any one adjustment might – and likely will be – imperfect, what does the consideration and adoption of these alternative approaches do for the decision-making process? What they represent are potential opportunities to improve, at least incrementally, on your understood objective.

Now, digging a little deeper into the mechanics of the process of figuring out how to go about packing your bag, how exactly did you go about evaluating the "goodness" of the solution you went with. Did you adopt a maximization approach, where you simply attempt to maximize the total number of items that you can fit into the bag? Was it a minimization approach, where you attempt to minimize the perceived cost, and risk, of relying either on checked baggage and/or parcel shipping to get your most important items home? Were you prioritizing in ways that weren't reflective of the size of the items being packed or the number of items being packed? If so, what were your priorities? And, perhaps most directly getting to your vision of "what the problem" was … which items did you end up NOT packing into your bag at all?

As you can certainly tell at this point in the breakdown, exactly how we choose to define what 'performance' means in the execution of a task is fundamental to the approach we adopt when tackling that task. The end point in the process defines what steps we'll take in its execution. Structure, of even the simplest architectural configuration or framing (e.g., in this case having a limited amount of space in your bag to pack for a trip) forces us to deliberately and explicitly confront inherent tradeoffs that exist in reality. It sharpens how we build out and develop a working definition of what 'performance' actually means in a given space, and might even wind up changing how performance is – or can be – defined at a given point in time. Structure also can systematically heighten our valuation of a given set of potentially available alternatives, in light of the (unfortunate!) reality that we can't always do (or, pack!) everything that we'd like to.

Note: Incidentally, packing problems including the knapsack problem, the strip packing problem, and the bin packing problem are staples of planning discussions and algorithm development in the operations management space across all levels of emphasis, from advanced undergraduate curriculum to on-site management. This framing encompasses everything from extreme, macro-level architecture reflective of how 400-meter-long modern container ships are loaded with 20,000-plus shipping containers to the extreme micro-level architecture of how ready-for-assembly furniture is packed into a single box. This sort of configurational task has also not surprisingly found its way into popular culture video games such as Nintendo's 'Professor Layton and the Diabolical Box', Getaway Entertainment 6 Pack's 'Stuffin the Briefcase', and even, in an abstract way Alexey Pajitnov's ever-popular – and extremely addictive – 'Tetris'. Incidentally, there is a way to get all 14 objects into the space depicted in Figure 3.2, with a little creativity, or perhaps the help of a computer (see Appendix A.1 for a solution).

Having completed this exercise, and having worked through various approaches to getting everything you wanted into the bag, you could very easily come away with the view that the inherent limits that structure imposes on the decision-making process generate more problems than benefits for both problem development and problem resolution. However, whether we like it or not, the world around us – at home and in our personal lives – is defined by structure. We are resource-constrained. We just do not have unlimited resources available to improve the things in our lives that are important to us, to our families or to our organizations.

We can choose to ignore this reality, but our bosses and clients, children and spouses, would very likely be pretty upset with the resulting 'solutions' we returned if we did ignore this reality. Because after all, absent a real connection to the realities that define the structure of a problem space, these wouldn't actually be real solutions at all. Approaching the development and resolution of a problem without paying appropriate homage to the structure of the space, and to the inherent constraints defining our degrees of operational freedom, would be like trying to complete a crossword puzzle but only knowing the number of characters in each correct answer, but not knowing how they crossed over each other in space.

You might start off thinking that you have a lot more flexibility coming up with answers to each clue. But, by the conclusion of the puzzle the probability that your answers will ultimately actually end up fitting together in a comprehensible way is next to zero. Lacking structure in such instances also prevents you from capitalizing on insights that would otherwise emerge from incremental advancements (e.g., the solutions to 1, 2, and 3 Down can't help you come up with solutions to 1 and 2 Across if the clues don't overlap). A lack of structure not only is likely to result in impractical solutions, its absence can also fundamentally slow down your progress.

1.2 The OUtCoMES Cycle: An Overview

Developing the best problem definitions, and ultimately effectuating successful problem resolution in exemplar projects, is of course a complex process, which really does go without saying. Adopting an ad hoc approach to problem definition fundamentally undermines the goal of effective problem resolution. Great problems and ultimately great projects ultimately emerge from great processes that capitalize on strong frameworks that provide an explicit demarcation of problem architecture and degrees of operational freedom. The approach we develop provides these channeled mechanics, and like any well-architected system encompasses both content and direction, stocks and flows. We refer to this approach, which we return to repeatedly throughout our discussion in the book, as the OUtCoMES Cycle (caps used deliberately; c.f. Bendoly 2020). The core elements of The Cycle are presented in Figure 1.3.

As the name suggests, the OUtCoMES Cycle (aka simply as 'The Cycle' in this book) encompasses an iterative feedback process. The Cycle provides

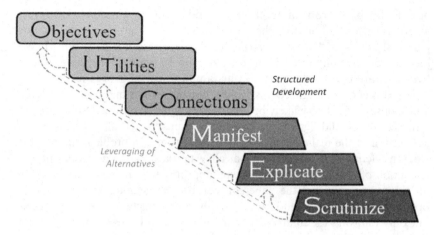

Figure 1.3 The OUtCoMES Cycle at 20,000 Feet.

structure to otherwise unstructured, or poorly structured, real-world problems. The Cycle is both a framework for problem-evolution documentation, as well as a process of discovery. The Cycle begins with the identification of core performance Objectives, critical levers (e.g., Utilities), Connections and salient limitations on managerial decision-making. Core performance objectives include, for example, improving key organizational metrics such as ROI, retention, sales growth, waste, carbon footprint mitigation, etc. Objectives are both organizationally relevant, as well as analytically tractability, and are thus responsive to the application of descriptive, predictive, and prescriptive analytics. Critical Utilities are the useable levers available for better describing, predicting, and/or prescribing functional changes. These levers can include anything from how much money being invested in training, to how and when to modify pricing, to the kinds of contract modifications made to secure future environmental criteria in supplier selection. Connections entail the specific forms through which Utilities impact Objectives, or interact through tradeoffs (e.g., due to fixed budgets) and synergies (e.g., additive virtuous circle benefits).

Applicable across a host of analytical tactics in model development and estimation, the Cycle helps guide exemplar project development, redevelopment, and, ultimately – ideally – actionable practice prescriptions derived from formal consideration of alternative structures of cause and effect. Strong analogies are present between the Cycle and well established, though insufficiently articulated flywheels, such as PDCA (Plan, Do, Check, Act; aka the Deming Cycle of continuous improvement) and DMAIC (Define, Measure, Analyze, Improve, Control; often used in 6-sigma-anchored project settings). Similarly processes are in fact fairly ubiquitous (c.f. Ken Watanabe's Problem Solving 101, 2009, p. 14). These processes also capture the critical principle of iterative re-examinations the Cycle promotes. However, they fall short in regards to specific recommendations for exactly 'how' to develop, and redevelop, the problems they are presumed to elucidate. The processes they promote also lack structural specificity or explicit specification of alternatives. While these more generalized processes may facilitate teams' performance, their use also provokes vulnerability and susceptibility to very serious, and common, traps including escalation of commitment, reinforcement bias, dissonance between engineering/analytical success measures vs. practical relevance and priorities.

Now, for those of you reading this book who also are fans of frameworks like the A3 and PDFA, the DMAIC, Design thinking, and the double diamond process – we want to take a moment to be clear about something before moving forward. We are, broadly speaking, fans as well. To be more precise, we are fans of any process or framework that can help people to come up with great solutions to great problems. We are also not dogmatic in the sense that we would never suggest that any one of these frameworks or approaches is the only, or the best, approach to structuring problems or deriving actionable recommendations or value-added intelligence around them. We believe strongly in the value of multiple perspectives. That is how we approach this entire process, and how each of us views The Cycle. As a process and a framework that is not merely an attempt to "replace", but rather to complement, venerable and useful tools that many of you are already very familiar with, and likely have used numerous times in the past in your own project work. This will become apparent as we draw on, and augment for more systematic consideration, several classic tactics such A3 documentation.

We delve into each of the elements of The Cycle, beginning with a discussion of how objectives that define problems are best developed and assessed (Chapter 2). We then discuss similar processes for evaluating the relevance of Utilities and Connections (Chapter 3). We explore the systematic assembly and subsequent re-configuration of these components into a coherent model form (**Manifest**) in Chapter 4, which is paired with a discussion of tactics to extract insight from these composite models (**Explicate**), with an emphasis on checks to face validity and practical relevance (**Scrutiny**).

🌓 *Disappointment in a Borderless Sandbox*

A few months ago, following a keynote at an industry conference, an audience member very frustrated by his own recent experiences shared an unfortunately all too familiar story. At his direction, a year earlier his firm had hired a highly skilled (and very expensive!) analytics team. A clear return on the investment had yet to materialize, and his own bosses were raising eyebrows. He was taken aback when asked, "What is analytics team trying to accomplish?" He answered, "Well, they're supposed to be finding ways for us to make more money ...' In response to the follow-up 'What actions were you hoping would come out of their work?' he replied 'Anything and everything ... "

Of course, he hadn't really meant "anything and everything", but his response was telling, and its terseness (and generality) encapsulated the ultimate root-cause of his intense frustration. It also hadn't had anything at all to do with the skillset of the new hires, who came from the very best programs in the country and had had extensive prior work experience. The core issue was that the team simply hadn't been given enough direction. The manager, despite years of experience, hadn't provided the analysts with enough problem structure to ensure yield. The team had been left to play in a sandbox without boundaries. Plenty of great tools, but the space in which discovery could be made was left far too wide open, and the absence of limits had fundamentally undermined the team's ability to return a useful outcome. Raised eyebrows all around ...

They had never been told – 'These are our performance priorities.', 'These are the things we can change.', 'These are the constraints we have to work with.' Basic directions and best discovery practices could have launched the team in the right direction from the get-go, generating immediate pay dirt, and eliminating second-guessing. Encouraged to engage in transparent and frank discussions of scope, intermediate outcomes and challenges, the frustrated audience member was intrigued, and eventually sought our help to establish a more formal set of guidelines for future efforts. The resulting structure made all the difference ... better late than never.

1.3 Valuing Your Toolbox, Not Just the Hammer

Importantly, the Cycle directly informs, and is informed by three broadly ubiquitous genres of data analysis: (1) Descriptive Analysis, (2) Predictive Analysis, and (3) Prescriptive Analysis. Described later in this section, these help answer What, Why and How questions such as "What" is the history/limitations of our objectives, "Why" might these have changed over time?, and "How" might additional changes further drive performance, subject to limitations.

Descriptive Analysis: Descriptive analysis helps clarify "what" we are dealing with in a problem. We might think of this as 'describing our local reference system' (see Section 1.1). That is, the most effective descriptive analysis informs us regarding the most central issues that we are contending with, and those factors that may influence or limit those issues. If the principle objective is profit maximization, for example, descriptive analysis elucidates relevant elements of the problem space relating to profits. These depictions could include summaries of current average annual or monthly profits, historical profit ranges over a given period, or the historical shape of the profit distribution over that same period. If the objective is developing more realistic predictions of demand, again, descriptive analysis informs aspects of the problem space bearing on this criterion. This framing might include for example whether demand could be most accurately described as homogenous across potential market segments, or as being more multimodal, with pockets of demand gravitating around different combinations of geographic and/or socioeconomic dimensions. These are important questions to ask because an understanding of the performance landscape is a critical point of departure for any efforts to explore it, anticipate its likely fluctuations, or indeed deliberately generate functional changes. It is critical to build understanding of what is 'possible' in an omniscient narrator sense, and where limits are likely to arise. It is critical to closely scrutinize all of the relevant pieces the pieces of the puzzle that encompass or define the problem space before trying to actually piece this space together. This can involve using tactics as basic as measures of centrality (e.g., mean, median, mode, etc.) and variation (e.g., range, standard deviation, confidence intervals, distribution fits, etc.). But, this can also involve more complex tactics such as exploratory K-means cluster analysis, or even more advanced tactics such as GMM (discussed in Chapter 5). Investments in descriptive analysis should be deliberate and explicitly focused on objectives and Utilities. Descriptive analysis should more heavily emphasize understanding and explication of descriptive details relating to objectives and Utilities from the Cycle than speculation relating to the exact form of their relationships with one another (which is the focus of predictive analysis).

Driving home this latter point, Figure 1.4 depicts the relative focus of Descriptive, Predictive, and Prescriptive analysis across the stages of the Cycle. For example, during the initial consideration of Objectives, descriptive analysis is critical. Close scrutiny of the robustness and practical applicability of solutions similarly benefits largely from descriptive considerations. We need to 'describe' the findings of analysis, shortcomings of that analysis, plans of action, and risk. Only then does it become possible to determine whether to go back to the drawing board or forge ahead in implementation and next-level projects. In contrast, outlining the nature of Connections, defining how Utilities (i.e., these levers for change) relate to Objectives chosen for pursuit, and explicating

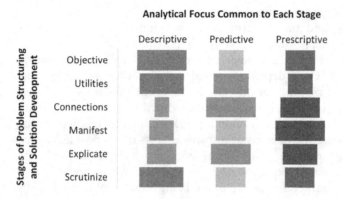

Figure 1.4 Relative Analytical Focus by Stage.

their empirical form (i.e., fitting parameters to that form), is typically the focus of predictive analytics.

Predictive Analysis: With Utilities, we also have an interest in description, though that interest chiefly emerges from the mechanical aim of advancing a particular Objective. Utilities are the levers used to help explain, anticipate and eventually guide changes in Objectives. Approaching these future changes realistically depends on knowledge of how far these levers have been pushed or pulled in the past, and how much further they might be pushed or pulled in the future. For example, understanding the history and consequences of past experimentation in adjustment of investments in quality is a critical first step informing current investments to yield net increases in profitability. If these investments had previously seen very little adjustment, is there a reason why? If these adjustments were higher in the past, but were recently reduced, again what underlies this pattern? Perhaps of greatest interest, are there historically positive associations between investments in quality and profitability growth, and have these changed? These questions require clarification of "why" in order to predictively combine objectives, Utilities and their Connections with one another. While Objectives remain of principle concern, predictive analysis tends to emphasize explication of the role played by Utilities and the form/nature of their associated Connections. Prediction[2] can be accomplished using an array of tactics, from basic linear regression to more sophisticated machine learning approaches for classification. These tactics facilitate understanding of not only why things may have happened in the past, but also provide guidance for movement toward achievement of future objectives. At this stage in the process, formally, the pieces of the puzzle are fitted together to develop a clearer picture of the maze of potential opportunities available. This process of clarification is what increases the probability of making sense of – and determining – the best path forward as regards determination of an appropriate objective.

Prescriptive Analysis: Beyond understanding and anticipation, prescriptive analysis advances this analytical process to the next – and arguably the most important – analytical phase. Prescriptive analysis explicitly emphasizes the operational "how", where the intelligence generated from descriptive and predictive analysis is used to formulate a systematic solution. While descriptive and predictive analysis are critical to piecing together the puzzle (i.e., structuring the problem), prescriptive analytics facilitates the development of tactics, options, and limitations necessary to navigate the twists and turns of the maze that emerges from this process (i.e., development of practical solutions). Prescriptive analysis incorporates a comprehensive view of our local reference system of focus (Section 1.1), accounting for the implications of multiple predicted consequences of actions, tradeoffs between actions, synergies between actions, potential risks and returns associated with a range of actions, and the fundamental constraints on decision making implied by organizational, legislative and physical laws. Because of the importance of developing a coherent systematic picture, in prescriptive analysis a substantial amount of time is devoted to articulating that picture. Here, an interdependent system of Connections from which a potential 'best solution' can be explicated is developed. Although optimization is the most common frame adopted in prescriptive analysis, optimization tactics can range from basic Simplex approaches to more complex approaches leveraging, for example, Genetic Algorithms. Formal discussion of these and other tactics is offered in Chapters 4 and 5.

The example built out in Figure 1.5 offers illustration of the interplay between Descriptive, Predictive, and Prescriptive analysis and the OUtCoMES Cycle. From top to bottom, we encounter tasks, approaches, applied objectives (the relationship between managerial and analytical objectives is discussed in Chapter 2), Utilities and Connections to these ends, structural details of manifested models, estimations from explication, and scrutiny of practical implications.

📟 *An Example with Token Data*

To help solidify the relationship between stages of the OUtCoMES Cycle and the three types of analysis described above, it can sometimes be useful to think about an example data scenario. Imagine that we have a data set with five fields:

	Descriptive	Predictive	Prescriptive
Analytical Task	Exploratory Grouping	Model Fitting	Decision Optimization
Basic Tactics →	K-Means Clustering	OLS Regression	Linear Programming
... and Other Options	GMM, Affinity-Prop	GLM, Time-series	GA, Swarm/Colony
Objectives *(Managerial)*	Classify Clientele or Organizations	Forecast Purchasing, Explain Performance	Plan Ad Placement, Allocate Investment
(Analytical)	Maximize ratio of between to within variation	Maximize goodness of fit metric (R-square, LL, etc.)	Maximize value of objective function, s.t. constraints
UTilities	Grouping variables	Predictors (IVs, controls)	'Decision' variables
COnnections	Number of groups	Linear, non-linear, lagged	←& Limits / budgets, xor
Manifest	Cluster size restrictions, hierarchical, distance definition	Weighted average, interactions, system of related equations etc.	Objective function and system of inequality constraints
Explicate	Assignment of observations to clusters	Assignment of values to variable coefficients	Assignment of optimal value to decisions
Scrutinize	Practical meaning or relevance to the groups?	Significant and robust predictive effects?	Prescribed course of action is viable / logical?

Note that confirmatory clustering would involve false positive and false negative rate considerations

Figure 1.5 Example Mappings of Analytical Elements to OUtCoMES Cycle Stages.

X1: A continuous measure that we believe someone has some control over (e.g., amount of money spent on a specific type of training)

X2: An interval measure that we believe someone might have some control over (e.g., number of employees assigned to a kind of task)

X3: Another interval measure that we don't have control over (like "year")

G: A nominal designation of grouping (e.g., task number 1, 2, 3, or 4; employee group a or b)

Y: A continuous measure that characterizes managerial, or organizational performance in this setting (e.g., revenue)

Universally, the best place to start is with an examination of the elements of the data set we are working with. It would be rash to assume a priori that any one of these data fields follows a particular statistical distribution, that the data don't need to be cleaned, or that the data doesn't contain errors that need to be reconciled against the source of the data (e.g., through discussions with the stakeholders who provided it in the first place). This first stage of examination falls within Descriptive

Analysis, as discussed earlier in the chapter. And, importantly, there are a host of questions that can be asked with this data set example.

Describe X and Y: What is a typical value of a numerical measure like X1, X2, X3, or Y in the data set? A sense of what kinds of values are typical can be obtained through measures like the mean or median, but these anchors provide only an initial snapshot of these data. We can also ask how can values of X1 can be summarized, or how Y values are distributed. Simple measures such as standard deviation can help achieve this end, if we understand the 'shape' of these distributions. Plotting these data in histogram form, or applying fit analyses to determine whether distributions are roughly Normal, Exponential, Uniform, etc. will help to get to that point. For other kinds of variables (e.g., nominal or even interval) we could also ask how many observations there are in each group. Again, simple count-based depictions (e.g., bar charts) can go a long way towards enhancing understanding of the data, but so to can exploratory clustering. That is, we can ask whether there are other ways to group (bin) things to capture important differences. Depending on the variables used in exploratory clustering, groupings distinct from the nominal classifications in G might become apparent. Ultimately, these questions ask "What is the current and historical state/nature of the data?" The value of Descriptive analyses can simply be in generating awareness of central data benchmarks, or the scope of observations, or flaws in the data that need to be addressed. However, the data can also be described by measures of fitness (e.g., measures of fit to distributions; measures of between/within variance in exploratory clustering, as in Figure 1.5's example). These measures are essentially Objectives that start at non-ideal, unknown levels. For example, consider a predictive model that can only account for 5% of the variation in a targeted outcome. While the ultimate aim is a stronger predictive model, the analytical Objective is to maximize understanding of variation so that predictions can be pinned down more effectively. Enhancing understanding of the drivers of the variation in that outcome obviously depends on analysis, and the specific choices made by analysts themselves (e.g., Utilities here include the choice of which variables to use in clustering; which variables to consider in distribution fitting). In an alternative example, the number of groups to form in clustering, or range of distributions to consider, impose limits on Connection between these Utilities and Objectives of our analysis. Insights gained here set the stage for Predictive and Prescriptive analyses.

Predict Y=f(β,X): Here the questions we are trying to answer include, for example: How much does a small change in X1 impact Y? In simple linear regression, this relationship could be captured in coefficient estimates (betas). We could also ask if X1 and X2 predict groups placement G? That would involve either confirmatory cluster analysis, or something like a neural network estimation. We could ask whether the impact of X1 on Y depends on X2 or G? In this case, estimates of size and significance would be assessed for interaction terms as predictors. Or we could use econometric approaches to determine how time (X3) figures in. These questions essentially ask "How can I anticipate future changes in Y as other things change?" All forms of predictive modeling have a variety of fit statistics, which we want to be as strong as possible, without overfitting. These are, in short, the analytical Objectives that we are keeping an eye on. The hope is that weights (e.g., betas) assigned to predictors work out. Such model parameters (e.g., betas) are the levers (Utilities) available in pursuit of model fit Objectives. The specific numerical structure of the estimated model relates to our assumptions about Connections (to be supported or otherwise through our efforts to build (Manifest), estimate (Explicate), and Scrutinize results.

Prescribe $\Delta X \rightarrow \Delta Y$: In prescriptive analysis, questions include: How much *should* we change X1 and X2 to impact Y? Or, to encourage future placement in one of the groups G? That is, rather than simply considering 'if' change is predictable … what is the best investment in pushing those levers (X1 and X2), given costs, returns and limits, to improve Y? Answering these questions requires a comprehensive examination of complementary effects and tradeoffs, limits to what we control and limits to how far outcomes can adjust. The result of this search for 'what to do' includes recommended changes to X1 and X2 (e.g., Utilities in this case), avoiding violations to rules present in our settings (Connections involving synergies and limits), towards real anticipated improvement in Y (Objective). The comprehensive models we develop in the Manifest stage create the backdrop for analytical searches for ideal (if not optimal) combinations of decisions during the Explicate stage. Scrutiny is then applied to assess whether the identified solution is workable in practice, or if additional modifications to assumptions and model form are needed.

1.4 Documenting Structure and Alternatives

Because documentation and the consideration of alternatives is so vital to effective engineering and analytics projects, the use of the OUtCoMES Cycle also emphasizes development of structured frameworks for documentation, brainstorming, and idea revision. Inspiration comes from the successful use of

A3 documentation in practice (e.g., one of several outstanding tools emerging from years of effort in rationalizing details and decision support within the Toyota Production System, c.f. Sobek and Smalley 2008). In support of the OUtCoMES Cycle, we augment the basic structure of traditional A3s by focusing on the importance of systematic clarity across the phases of model development. The result is the Systems-oriented A3 (S-A3). The S-A3 framework represents a substantive advance beyond venerable, though limited, continuous improvement documentation tools. While traditional A3 forms (structured 11' × 17' documents) are often used to convey a linear, coherent flow of logical and empirically grounded reasoning, the S-A3 is explicitly designed to function as a "living document." As an explicitly designed complement to the OUtCoMES Cycle, the S-A3 provides a documentation structure for outlining alternative objectives, as well as alternative views of how these objectives might be pursued. The S-A3 also facilitates development of extremely nuanced understanding and explication of possible cause-and-effect relationships, as well as practical limitations of both performance outcomes and putative performance levers. While subsequent chapters delve into each section of the S-A3 documentation in detail as we discuss the OUtCoMES Cycle, an overview of the S-A3 is provided in Figure 1.6 for preliminary consideration.

In the chapters that follow, we move from rich consideration of how to initiate projects to discussion of how best to seek out their solutions, with an emphasis on continuous learning and revision. In Chapter 2, we

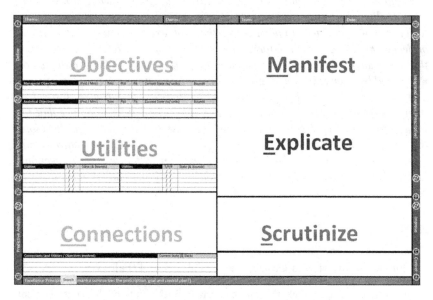

Figure 1.6 Overview of Systems-Oriented A3 (S-A3) for OUtCoMES Documentation.

Note: More detailed digital template to be discussed and available at: www.masteringdiscovery.com

address the all-important question – 'what's the problem?', scrutinizing various alternative responses that can emerge. In Chapter 3, we examine elements that can help to further define options for greater specification of Objectives, and that can ultimately guide action (Utility and Connections). Chapter 4 explores the process of building sufficiently comprehensive reference models for analytical processing (e.g., predictive estimation, prescription extraction, etc.), and develops expectations regarding the nature of solutions, the scrutiny applied to these, and the importance of capitalizing on alternatives established in earlier stages. The remainder of the book shifts gears to focus more on the coordination of combined tactics, managing the roles and ownership of project team members, facilitating the sharing of data and ideas, and how to span functional and organizational boundaries to maximize the value of projects through joint problem identification and solution development.

Practitioner Recap

The word 'structure' can sound a lot like 'micromanagement' to many analysts and engineers, and the idea of developing alternatives before things go wrong can seem like a waste of time. Contingency planning is often only given lip-service, and more often than not we see project teams scrambling to come up with a Plan B only after things have been going downhill for a while.

In reality, the right structure can inspire critical thinking and equip teams with the agility they need to pivot before they sink excess resources into a losing project proposition. Structure forces the consideration of priorities and increases the odds that the best paths are taken in projects. It drives effective discussion within and between stakeholder groups. And, if alternatives are outlined a priori, that discussion is also less likely to register barriers in advance, instead of running hard into them. It's more likely to yield compromise and ensure that the value coming out of projects is maximized. Because of that, documentation along the lines of the OUtCoMES Cycle's structured framework, encouraging brainstorming around alternatives, charting current to future state projections, and facilitating revision at each stage, regularly gets groups and organizations to performance heights they wouldn't be able to reach otherwise.

Notes

1 Popularly attributed to Albert Einstein. Articulated by John Archibald Wheeler in Interview in Cosmic Search, Vol. 1, No. 4 (Fall 1979).
2 While some prefer to distinguish between Predictive and Diagnostic analytics, the differences are largely that of perspective. Both aim to explain past, current and/or future variation. Their explanatory missions distinguish them from descriptive aims or objectively prescriptive ones.

Chapter 2

Picking the Right Problems

To get us thinking about the first few steps needed for the task at hand, let's once again consider a few motivating quotations from popular culture.

The scientific mind does not so much provide the right answers as ask the right questions.
 – Claude Lévi-Strauss, The Raw and the Cooked, 1964

It is not the answer that enlightens, but the question.
 – Eugène Ionesco, Découvertes, 1969

Good questions outrank easy answers.
 – Paul Samuelson and Kate Crowley, The Collected Scientific Papers,
 Volume 5, 1986

So, the question is … where do we start? At this point, we've started to build out the principles undergirding the framework and process for developing and evolving projects we call the OUtCoMES Cycle (or more colloquially just 'The Cycle'). The Cycle – from here on in the book – is an approach that can be used to increase the likelihood of generating strong returns on project investments. But, how does the Cycle work specifically? How do we use this approach?

In order to have any hope of generating value using the Cycle, the first step must be identification of what we're trying to accomplish. And, to do this we have to accept two simultaneous – dissonant – realities. The first is that precision in detailing where we want to end up is tremendously important. The second is that this specification is overwhelmingly likely to be imperfect, given that the landscape in which the project is embedded will only unfold over time. Ultimately, it may turn out that there are far better targets toward which available project resources could be focused.

DOI: 10.4324/9781003427650-4

Time and effort will tell. However, from the outset, projects enjoy great benefits from detailing what we *think* we'd like to accomplish; what kinds of questions we *believe* we are trying to answer; what sorts of problems we *feel* we are trying to tackle. This is true despite the fact that initial formulations of the 'right problem' may need to be replaced with something entirely different as the evidence unfolds.

The first, and foundational, stage in The Cycle, appropriately, is utilizing current understanding of the context to establish key Objectives. The use of the plural form, Objectives, is deliberate.

Objectives, in the OUtCoMES Cycle, describe measurable outcomes that serve as both the motivating force catalyzing project investments, and around which local reference system details orbit, as well as definitive yardsticks by which project success is ultimately measured a posteriori.

---

bjectives describe measurable outcomes that matter to the organizational, operational, and strategic context, and reflect how well the associated analysis performs. In his seminal discussion of 'Management by Objectives' (1954), Peter Drucker argued that professional outcomes are best advanced through the use of Objectives that align interests with justifiable actions, while Payne et al. (1999) warned against the pursuit of poorly substantiated Objectives. These cautions emerge from the reality that weak evidence in support of efforts to advance real-world performance fundamentally increases risk. Ultimately, the most likely outcome of efforts unsupported by relevant evidence is additional cost, not enhanced performance.

Thus, the principal focus of this chapter is understanding distinctions among Objectives, and approaches to identifying, contrasting, and prioritizing them. We start with a discussion of general classifications that facilitate identification, as well as several prototypical Objectives. We then break down the attributes of alternative candidate Objectives, contrasting these and setting up prioritization.

2.1 Forms of Objectives

While Objectives can certainly be distinguished using a broad range of criteria, two of the most important distinctions lie between motivation (Managerial) vs. evidence (Analytical), and between product (Fundamental) vs. process (Means).

2.1.1 Managerial versus Analytical Objectives

Not surprising to anyone reading this book, projects are commonly motivated by directives handed down directly by higher-level supervisors, stakeholders, or clients. They also may be motivated by needs identified by project managers or even individual employees. These motivational, **Managerial Objectives** typically reflect the desire to improve a measurable performance metric (e.g., profit, speed, quality, retention, capture, efficiency, etc.), or reduce undesirable outcomes (e.g., cost, service errors, waste, customer dissatisfaction, injury, etc.). In these cases, the benchmarks against which improvements are compared might simply be the current state of these metrics. This then begs the question, does the project offer a path toward increasing efficiency, or decreasing injury relative to historic levels, or merely to what's been witnessed most recently? Managerial Objectives also can, of course, be of the 'make' or 'adopt' variety, including for example new product development, third-party system implementation, or customer segmentation. In this case, there may be no obvious benchmarks as this may be the first time the organization has implemented a third-party system or classified its customers. While the motivation and perceived value of the project derives from replacing 'nothing' with 'something', there may often be at least implicit benchmarks: the returns from a new product relative to the last one developed, or the net cost reductions generated through system implementation.

It is important to be clear that accomplishment of Managerial Objectives is not equivalent to the project work leading to these returns. Managerial Objectives are typically lagging indicators of the effectiveness of tactics applied during the project, or of how well a project is going at any particular stage. Project teams, engineers, and analysts also have other barometers for evaluating tactical effectiveness. Measurements that emerge from testing and simulations in engineering projects can facilitate evaluation of individual design choices. Comparing these choices to existing design requirement benchmarks is central to modern AI-supported design prediction and optimization processes (e.g., generative design).

From a broader analytics perspective, regardless of data source, rigorously generated descriptive, predictive, and prescriptive evidence speak to a different category of Objectives: Analytical Objectives. We want to predict the impact of a change in a product feature, the impact of a staffing increase, the impact of the choice of a specific system module on Managerial Objectives ... because, ultimately, we want to know what can done to actually increase profits, reduce injuries, diminish carbon footprint etc. We want to come up with an optimized plan for resource deployment to advance Managerial Objectives because there are so many options, and limitations, bearing on these resources. The successful development of descriptive, predictive, and prescriptive models (aka **Analytical Objectives**) is fundamental to accomplishing Managerial Objectives, and importantly Managerial Objectives also can be tracked far more continuously and directly by project leaders and teams than Analytical Objectives.

Distinctions between Analytical and Managerial Objectives, across all three genres of analysis, are integral elements of the Cycle (as seen in Figure 1.5). Identification of parallels between Objectives of relevance to non-analytical stakeholders (Managerial) and those bearing on the minutia of process work (Analytical), also facilitates communication across professional dialect fault lines; a common team challenge (Thompson 2020; Goldman and Taylor 2023).

2.1.2 Fundamental versus Means Objectives

Clearly, neither Managerial nor Analytical Objectives are either easily or spontaneously achieved. Set up and preparation is always likely to be extensive. In order to achieve great things, more often than not more mundane preparations (to some) need to be competed. As any experienced project manager is likely to readily acknowledge, precedence is critical. And, as any experienced analyst or engineer will tell you, a lot goes into the development of a solid and robust predictive model. Moreover, optimization, often relying on knowledge that emerges from such predictions, can involve systematic considerations so complex that the search space can't be exhaustively mapped. In these instances, savvy heuristics and computational approaches become core to the tool set. And, just as complex managerial goals must be pursued in a stepwise fashion, individual analytical steps require stepping-stones to final prediction and prescription orbiting their own Objectives. This is how it becomes possible get to where we ultimately want to go analytically, and managerially. Accordingly, precedence and hierarchy are determined by gauging these stepping-stone efforts by what is called **Means Objectives**, adopting the terminology of Bond et al. (2008), in contrast with our higher-level **Fundamental Objectives**. Doing so provides further guidance in the prioritization of tasks, which also is subject to change as evidence emerges and priorities evolve. What had been a Fundamental Objective at a given point in the Cycle later can become a Means Objective.

Of course, introducing new ways to think about how project success is gauged can be an adjustment. On the other hand, it seldom hurts to at least consider an alternative vantage point. From our consultancy, Fundamental/Means designations coupled with Managerial/Analytical designations can be tremendously effective helping teams organize their efforts around discovery and value-adds. Because these designations are likely to be new to some readers, it can be valuable to think through an example of how Means and Fundamental designations intersect with Managerial and Analytical designations for determination of project Objectives.

🌑 *An Objective Roadmap*

An interesting recent project reflecting this framing involved two companies confronting a shared opportunity. The first was an original equipment manufacturer with a massive industrial footprint (the 'the OEM'). The second was an equally well-established world-wide service provider ('the Owner'). The opportunity they faced involved the use of modern sensor technology that would allow them to track the condition of various features of the equipment purchased by the Owner from the OEM. Realizing of the potential of this IoT (internet of things) opportunity depended on a several conditions being met. First, decisions regarding how many sensors to embed in the equipment, and how to receive and collect signals, had to be determined. Although these infrastructural builds fell largely within the purview of the OEM, they would hardly be interested in such expenditures if the data collected would be of no use. That, in turn, required a consideration of clients such as the Owner. The Owner certainly wanted to avoid critical and costly equipment failures, but it also wanted to avoid excessive additional costs. These could include price increases in equipment (e.g., folding in the sensor and infrastructure investments at the OEM), subscriptions of a sort to the data collected by the OEM, equipping and training personnel (e.g., direct users; technicians) in the use of sensor signal data, and excessive additional downtime associated with, for example, preventative maintenance. However, it would be difficult to justify an investment if it weren't possible to identify a combination of these factors that incurred costs lower than the marginal benefits associated with the sensor technology (i.e., a solution enabling the technology to yield positive returns for both sides). What was needed were analytical projections, derivations of potentially return-maximizing approaches in the form of policy and technical investment adjustments, and of course no small degree of hope that the net return projected from these approaches would ultimately be convincingly positive.

Getting to that final point involved a large number of intermediate steps. It required bringing together individuals with various technical and managerial backgrounds, and the organization (and in some cases new collection) of data to serve as the foundation for analysis. It required analysis describing the landscape, bounds on measures,

Table 2.1 Coordinated Objectives in with Policy and Investment Utilities

Objective Level	Objective Facing		Primary Genre of Methods
	Managerial	*Analytical*	
Fundamental-A	Advance long-term returns on equipment through policy and investments	Optimize (maximize) value/unit-time projections, by computationally comparing policies and investment options w.r.t. impacts modelled	Prescriptive
⇧	⇧	⇧	⇧
Means-A *(& Fundamental-B)*	Anticipate the potential impact of changes in work / data policy and investments	Specify predictive capability of models and explanatory contextual, policy and investment variable (i.e., detail R^2, LL, F1 score, p-levels)	Predictive
⇧	⇧	⇧	⇧
Means-B1	Adequately describe user and technician response to signals	Summarize user compliance, extra-role care, technician error, statistical power by time and context	Descriptive
⇧	⇧	⇧	⇧
Means-B2	Establish reliability and validity in data collection system	Summarize data feed latency and omissions, via centrality and distributions by time and context	Descriptive

descriptions of the ways in which events occurred in time (e.g., distributions), differences between contexts, costs associated with past, current, and future investment decisions. It required model development; e.g., first predictive and subsequently prescriptive, leveraging those predictions. And critically, it required systematic evaluations along each of these steps to build confidence in the analysis and the practical insights that emerged. Table 2.1 depicts the critical Managerial and Analytical Objectives pursued in this setting, with FOs at the top and supporting Means Objectives below. The pursuit of Objectives at the bottom were critical to the advancement of Objectives at the top.

Mapping the inter-related nature of these Objectives provided justification for the time spent and resources allocated towards analysis. Mapping these interrelationships also provided a structured, coordinated path informing a stepwise approach for that analysis, and also set expectations for the timing and nature of managerial deliverables. Breaking higher-level goals into a series of smaller goals also had the benefit of increasing the potential frequency of 'wins' across stakeholders, which in turn enhanced intrinsic motivation.

It should also be noted that as evidence evolved, so did focus. That was of course to be expected, but the mapping of Objectives helped anticipate those shifts. Near the end of the project, policy, and technology investment decisions played a critical role in both prediction and prescription. We refer to these as *Utilities* given their role as change levers (discussed further in Chapter 3). However, they were not the only relevant factors, and were not in principle focus at earlier, ground-setting project stages. Aspects of the context that were not seen as factors that could be directly controlled (e.g., shifts in demand), or which were not subject to change (e.g., the role of the equipment in the business), nevertheless represented critical potential sources of performance variation, and had an impact on key decisions. Insight bearing on the setting landscape was crucial for understanding how effective changes could be anticipated and installed. These early descriptive steps may not have been the sexiest aspects of the project, but the rest of the work simply could not have proceeded without them.

2.2 Formulating Candidate Objectives

The example demonstrates how these firms adopted a coordinated, stepwise approach to develop critical Objectives, highlighting both managerial relevance and analytical performance metrics. But how were those Objectives identified in the first place? Objectives are often handed down to analysts, engineers, and project teams from external stakeholders, higher-level managers, or clients. But sometimes objectives have a purely organic inception. How do these emerge? Even if Analytical Objectives naturally complement clear Managerial Objectives, and even if Managerial Objectives naturally derive from FOs ... what about those end-game **Fundamental Managerial Objectives (FMOs)**? From among all of the options potentially available in complex systems, how do we identify the best of these?

Let's start with a retrospective thought exercise, and a little brainstorming ...

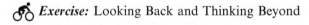 *Exercise:* Looking Back and Thinking Beyond

For this exercise, first try to think back on one of the last projects that you were involved in (e.g., new product development, technology implementation, organizational change efforts, expansion planning, process improvement, etc.). The project can have been at your own organization, or in another work setting. Take a minute, and try to select a project where you had a strong understanding of the motivation for the project, the approach taken during the project, and results from the work.

A few action steps will help to pin down some of these details.

[Action Step 1 – Setting the Stage]
Give a brief description of the project, in Table 2.2. Try to describe the project in a sentence or two.

Table 2.2 Recalling Your Project

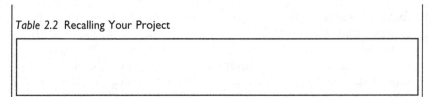

   This description helps primes your thinking, and may remind you of specifics and particulars of the effort. If you have difficulty building out the description, select another project to anchor on that you may be more familiar with. Otherwise, let's move on to the next step.

[Action Step 2 – Retrospective on Performance Foci]
Outline all of the relevant Objectives you feel were (or could have been) worth pursuing. Think in terms of Managerial – not Analytical – Objectives i.e., think of the 'ideal' outcomes that the non-analysts/non-engineers were discussing for the project. Focus on Fundamental Managerial Objectives (FMOs), even if these may have differed from the outcomes favored by the analysts/engineers involved. That is, prioritize the Managerial Objectives that were more endgame focused and less intermediate. Although we'll limit these to five, try to come up with at least two or three.

[Action Step 3a – Considering Alternatives ☐]
Let's do some comparisons. Below is a fairly extensive list of common FMOs. Check all of the boxes (☐ → ☑) that you think might also have been relevant, even if these were not formally discussed in your past project. The Managerial Objectives you identified above might also be listed here, and you should check all of those as well. Don't make any marks in either the ○'s or ◇'s yet!

[Action Step 3b – Mapping Commonality ○]
Even if the exact wording was somewhat different, did any of the FMOs you listed in Action Step 2 (Table 2.3) appear in Table 2.4? If so, it is important to explicitly acknowledge this overlap. Map the FMOs from Action Step 2 to this list. For example, if one of the FMOs you recalled was 'Increase Efficiency', put a check ✓ in the circle ○ to the left of that item in Table 2.4.

   If some of the FMOs you recalled in Action Step 2 did not map to the list in Table 2.4, write them into the empty cells at the bottom of Table 2.4 with the light grey geometric icons. IF you add items to the bottom of the table, just for consistency, check off the associated gray squares ☐ as well.

Table 2.3 Fundamental Managerial Objectives (FMOs) Recalled

i.	
ii.	
iii.	
iv.	
v.	

Table 2.4 Common Managerial Objectives

☐ O ◇ Raise Revenue	☐ O ◇ Market Capture
☐ O ◇ Reduce Costs	☐ O ◇ Strengthen Brand
☐ O ◇ Increase ROI	☐ O ◇ Challenge Competition
☐ O ◇ Increase Sales	☐ O ◇ Improve Leadership
☐ O ◇ Improve Pricing	☐ O ◇ Improve Training
☐ O ◇ Reduce Risk	☐ O ◇ Improve Compliance
☐ O ◇ Build Sustainability	☐ O ◇ Increase Quality
☐ O ◇ Improve Distribution	☐ O ◇ Customer Experience ↑
☐ O ◇ Increase Efficiency	☐ O ◇ Customer Relations ↑
☐ O ◇ Increase Productivity	☐ O ◇ Customer Retention ↑
☐ O ◇ Streamline Processes	☐ O ◇ Corporate Culture ↑
☐ O ◇ Build Capabilities	☐ O ◇ Employee Experience ↑
☐ O ◇ Increase Knowledge	☐ O ◇ Employee Retention ↑
☐ O ◇ Increase Innovation	☐ O ◇ Shareholder Value ↑
☐ O ◇	☐ O ◇
☐ O ◇	☐ O ◇
☐ O ◇	☐ O ◇

[Action Step 3c – Prioritizing Alternative Objectives ◇]
Drilling further into the FMOs with a ☑ (or ☑ in gray), for only those box-checked FMOs, use the diamonds ◇ to rate how important these FMOs were to gauging project success. As in Action Step 2, rate these objectives to reflect the views of those advocating for the project, albeit not necessarily the views of the engineers/analysts. For simplicity, use a scale ranging from 1 to 5 where 1 = most important and 5 = least important. Do not rate any unchecked FMOs, which we'll label as 'not important,' earning a rating of 6 or higher.

[Action Step 4 – Boiling Choices Down]
With your FMO ratings now documented, we'll build out a summary which you'll do by filling out the summary table, below (Table 2.5). For example, in the top row, how many of the FMOs from Action Step **3a** have a ✓ in both the square and circle? What was the

Table 2.5 Tabulating Commonality and Novelty

	Count	Min rating	Max rating
Checked items in Fig. 2.4 that Map to originals in 2.3			
Checked items that Didn't Map ☑○			
Original items added to Bottom of list			

minimum and maximum rating from that subset of FMOs? For the second row, how many FMOs only had a check in the square (☑ ○), and how were those FMOs rated? And finally, how many items from Action Step 2 could not be clearly mapped into the larger list from Table 2.4, and had to be added to the blanks at the bottom of the table? How did you rate these added-FMOs?

[Action Step 5 – Reflection]
What was the count of original items from Table 2.3 that easily mapped to the 'common list' of FMOs from Table 2.4? How many of the 'common' FMOs from Table 2.4 could have been important to the project, but just were not given much attention by key stakeholders? And, in retrospect, how important might those FMOs have been to the project? The answers to these questions are bound to vary, even across similar project contexts. But, when professionals complete this exercise, results similar to those in Figure 2.1 tend to emerge. As a rule, informed stakeholders commonly recognize the absence of multiple, critical FMOS from their recent projects, as well as a large number of Managerial

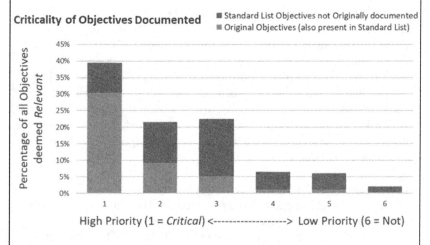

Figure 2.1 Typical Omission of Relevant FMOs in Project Retrospection.

Objectives that have at least some non-trivial level of importance (e.g., ratings of at least 3).

These results are similar to those reported by Bond et al. (2008), noted, "In three empirical studies, participants consistently omitted nearly half of the objectives that they later identified as personally relevant."

What is the key takeaway from this exercise? A common reaction is often a great deal of doubt about what actually motivates many real-world projects. Another very common reaction is cynicism as to the motives of key gatekeepers. But we can strive to do better, and the Cycle offers an approach that can move us closer to that end. Critically, while this was a retrospective exercise, where the focus was on what 'were' the FMOs of focus, or what 'could have been' other FMOs ... A more effective approach to scrutinizing Objectives should clearly occur in advance of projects. The selection of FMOs should always be done with a set of alternative, candidate Objectives in mind. The evaluation and prioritization of those candidate Objectives, like everything else, benefits from a structured approach.

2.3 Evaluating and Prioritizing

What often makes it very difficult to come up with the best Objectives (i.e., the 'best kinds of problems') is exactly the same thing that funnels certain projects toward poor solutions. The process for selecting Objectives is informed, in fundamental ways, by often deep (but not necessarily broad) professional experience (Thompson 2020). Accelerated by pervasive organizational mantras and industry keywords (e.g., 'Maximize Profit!'), these influences can have an outsized effect on the identification of FMOs, as well as those FMOs handed down by others. These influences generate disregard for the value of certain Means Objectives and can impact determination of Analytical Objectives. This can in turn lead to escalation of commitment to less ideal Objectives, especially when other options (e.g., potential Means; alternative FMOs) have not been considered in advance, and as a matter of best practice.

What turns out to be truly helpful to avoid this under-performance trap is a structured approach to scrutinizing Objectives, particularly with comparison to other less obvious options. At a bare minimum, this can facilitate determination of the viability (or non-viability) of otherwise perfunctory Objectives. It can also prompt development of more contextually nuanced Objectives. Fortunately, this front-end scrutiny builds from a long history of options assessment. The following example illustrates insight from more than two and a half centuries ago:

Dear Sir,

   When these difficult Cases occur, ... my Way is, to divide half a Sheet of Paper by a Line into two Columns, writing over the one Pro, and over the other Con. ... When I have thus got them all together in one View, I endeavor to estimate their respective Weights ... . I have found great Advantage from this kind of Equation, in what may be called ... Prudential Algebra.

– Benjamin Franklin, in a letter to Dr. Joseph Priestly,
discoverer of oxygen, London, September 19, 1772

Taking the time to outline Pros and Cons (or appeal vs. lack thereof) for any given 'candidate' Objective presumes at least two realities. The first is the existence of one or more alternate candidate Objectives (i.e., Objectives that might contend for position as an FMO). Having multiple options allows for scrutiny and comparisons to help determine project focus. The second is the existence of dimensions along which candidate Objectives can be evaluated for comparison purposes. For the first, if nothing comes immediately to mind, ideas from historical cases including 'common' lists (e.g., Table 2.4) can be drawn upon. We might also suggest a simple brainstorming activity such as Problem Statement (Why/Who/When/How) activity often applied in design thinking exercises (cf. Pg 49, The Design Thinking Toolbox, of Lewrick, Link and Leifer 2020). For the second, while there are a host of dimensions one might dream up for evaluation purposes, we have found a few to be particularly useful: Transparency, Plasticity, and Fit.

2.3.1 Transparency as an Objective Criterion

You may have heard the saying (or some variant) that 'You can't manage what you can't measure'. This message is often attributed to Peter Drucker, but almost as frequently to W. Edward Deming. While both certainly made an indelible mark on the way that Business (with a capital 'B') works to achieve improved performance, this idea isn't without its detractors. If you search the internet, you'll find those who believe that managing things does not in fact require their measurement. While that can certainly be true, it largely applies in two cases. The first encompasses scenarios where operations predominantly advance in a steady way, or where managers can delegate work such that functions essentially run themselves. If you don't really need to manage anything, you can probably get away without measuring anything. Of course, it isn't clear how long a manager in this situation might be able to coast in this way.

   The second encompasses scenarios where management doesn't know how to use measurement. Unfortunately, there are an overwhelming multitude of practicing managers with no background whatsoever in the analysis of evidence or data. No matter how well things are measured, these measurements will ultimately have very little impact on the way these managers do their work. In complex settings, data can be costly to collect, curate, and share; and if data are unlikely to actually be used, motivation to collect it in the first place is likely to be marginal.

However, for argument's sake let's assume there are managers, project leaders, or even intrepid engineers and analysts willing to admit they don't know everything in advance, and that evidence is critical to generating relevant, novel impacts in complex settings. If you're reading this book, you likely fall into one of these categories. Certainly, then, measurement is crucial to the sort of management you are interested in advancing. Transparency, as a measure of the viability of an FMO, captures this reality. In order to rigorously evaluate the potential impact and value of a project, that potential has to be benchmarked against current realities. This can't easily be done without transparency, access to measurement, and data visibility. Rigorous assessment of the potential trajectory or evolution of an outcome over time, either based on the organization's efforts or on factors outside of their control, simply isn't available without transparency of past data and on-going measurement efforts. And, neither viable prescription of a path forward, nor charting progress toward that goal are available without a similar level and breadth of transparency.

Viewed through any lens, transparency is an exceptionally important criterion for assessing the viability of Objectives, and thus a key metric on which candidate Objectives can be prioritized. When evaluating the transparency of any given candidate Objective, it is important to start with questions such as those below:

- In what unit is the candidate Objective measured? (e.g., dollars? days? a count? a percentage?)

  If this basic (and fundamental) question can't be answered, this is a serious problem. It is hard to imagine using any numbers associated with a candidate's Objective if you don't know what they are numbers 'of'.
- What is the current level of the candidate Objective, and/or what is a typical historical measure of the candidate Objective? (i.e., the 'current state' of the candidate Objective, or the candidate Objective's current average or median).

  In many cases, candidate Objectives measured in a single time period can be misleading. For example, the number of units that happen to be in stock on any given day is a poor measure of how much inventory is 'typically' available, let alone the cost-effectiveness of the inventory policy. But if it isn't possible to pin down recent, central measures of a candidate Objective (e.g., the median level of inventory for a particular

SKU measured over a year), the probability of generating meaningful analysis that can generate improvements is pretty low.

- How often is data collected on the candidate Objective?

Although an answer to this question may naturally emerge from the first two questions, it might not. Regardless, frequency is an important consideration. If you hope to explain past variance, anticipate the future, and ultimately develop effective prescriptions demonstrating impact ... anchoring this process against data collected once every few years poses a serious challenge.

- By whom, and under what circumstances, are data collected?

Knowing what data are collected, and how often, is crucial. But knowing the context in which data are collected and the principals central to the process can be critical as well. Data can be collected carelessly, or by stakeholders with vested interests and biases that lean in a particular direction. Some of the data may only be collected when certain circumstances are operative (e.g., failures; injuries), or it may be collected differently under different conditions. Having faith in analysis relevant for prioritizing candidate Objectives depends on having faith in the evidence on which the analysis is based.

Looking back on the Objectives you evaluated in Action Step 3a, using a scale ranging from 1 (1 = Very Low) to 5 (5 = Very High), how would you rate the transparency of each of these candidate Objectives at the time of the project?

2.3.2 Plasticity as an Objective Criterion

Consideration of some of the issues orbiting candidate Objective transparency provides insight into the second critical selection criterion: plasticity. If relevant data are being collected regularly, and if variance in these data is observable, there is some foundation for developing a plan of action to deliberately achieve desired changes, which leads to the following diagnostic questions.

- Is there variance in the data describing the candidate Objective which suggests it can reach more desirable levels? Or, is it already at – or close to – a functional limit?

Variance and control, of course, are not the same thing. It is possible that the vast majority of variance in a candidate Objective cannot be explained by other available sources of data. It is also possible that, while considerable variance in a candidate Objective is explainable the majority of explanatory sources are exogenous; i.e., not subject to direct manipulation.

If there are no degrees of freedom for additional improvements in a given candidate Objective, this can make that option untenable. If performance is already at 99% of its theoretical maximum, is performance the right Objective to focus on? Given common exponential increases in the cost of advancement, are there likely to be more fruitful opportunities? These considerations reflect classic framing offered in the 'Theory of Constraints' by Cox and Goldratt (1986). Is it clear how much the full potential of a system can benefit from a given candidate Objective, as opposed to others that might more severely bind/constrain the bigger performance picture? Are those latter issues part of the candidate set at this point? Bounds and limitations are fundamental aspects of a system, and like any other set of organizational rules or physical laws, they are part of the core structure of the best problems and exemplar projects. While understanding constraints provides guidance in problem develop-ment and resolution, ignoring them leads to misspecification and impractical solutions.

Observable variance, coupled with frequent data collection, increases the opportunity to identify the most impactful set of causes and effects, and also reduces the likelihood of simply throwing water against a wall. If little change is observable, perhaps there is a reason. If outcomes seem to have peaked, there may be little opportunity to gain value from additional project investments. However, if sufficient variance is present, especially over recent history (e.g., the last six months or year), the following questions emerge as a natural extension.

- Is ostensibly related data available, captured with an equivalent level of frequency, to help explain that variance? And is it realistic to control any of the predictors of that variance?

This question addresses the potential for prediction and prescription. If increasing predictability in the Objective, in order to be forewarned of – and be in position to plan for – risk is the principle (or only) aim of data collection, it may not be critical that strong predictors (e.g., time, weather, the price of tea in Boston, etc.) are uncontrollable. However, if the principle aim of data collection is development of predictive models as part of a larger prescriptive effort, where predictors themselves are also potential targets for change, the capacity for prescriptive modeling is paramount. If the aim is to make a difference, it is important to identify candidate Objectives for which there is a realistic opportunity to control one.

Looking back on the FMOs you evaluated in Action Step 3a, using a scale ranging from 1 (1 = Very Unlikely) to 5 (5 = Very Likely), what do you believe was the chance, at the time, for moving the need in the direction that key stakeholders hoped it would be moved in the project?

2.3.3 Fit as an Objective Criterion

The third major criterion relates to whether pursuit of any given candidate Objective is in fact aligned with all of the other moving parts in the system. Project investments, prescribed changes, and even the implications of predictive and descriptive analyses do not exist in a vacuum. They involve commitments of time and resources that might otherwise be allocated differently. The pursuit of any given candidate Objective could be at odds with other priorities. Working on engineering and analytics projects for an organization, internally or for external clients, has to be done in consideration of the broader (internal and external) contexts in which the project and organization are embedded, leading to questions such as the following:

• Does the candidate Objective align well with what is currently happening in the organization? Does work in the service of the candidate Objective align well with what key stakeholders are in the position to tackle effectively, even with the best data available?

This can be tricky to assess, and involves looking at more than data and designs. It requires consideration of what encompasses the entire problem reference frame. This can include stakeholders above the project leader, or above the clients. The overall strategy of the organization or other key players. What these players are currently dealing with, beyond what might be happening in the project. If an organization is facing significant dynamics, such as recovering from losses, or striving to maintain a leading position, or integrating or shedding significant aspects of the business or technology, additional efforts should also probably be delegated to these efforts. Supporting these efforts where possible, in these *firefighting* or *pacesetting* situations. However, if the encompassing context is not defined by such dynamics, then windows for more radical out-of-the box candidate Objective pursuits, and stretch projects, may have more upside viability (Figure 2.2). Of course this isn't a hard and fast rule. Sometimes out-of-the box thinking in projects is exactly what firefighting and pacesetting efforts need. Organizations don't tend to support efforts that appear to pursue Objectives residing off-tangent, and that in itself can be a major barrier to transparency and plasticity.

Looking back on the candidate Objectives you evaluated in Action Step 3a, using a scale ranging from 1 (1 = Very Poor Fit) to 5 (5 = Very Strong Fit), how well do you believe these aligned with broader organizational interests at the time? Do you think that some of these candidate Objectives might have had more support if they aligned better?

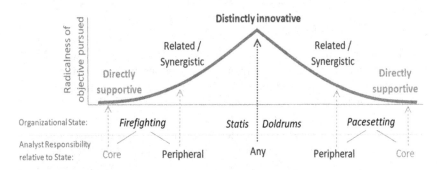

Figure 2.2 Organizational State and Radical/Incremental Nature of Objectives.

It is worth noting that Design Thinking discussion offer some analogous concepts: Feasibility ("technical implementability"[1]), Viability ("profitable") and Desirability (meeting "needs of customers/users"). Viability and Feasibility are subset-concepts implied by Transparency and Plasticity (certainly "profitable" is not always core to the changes we seek in Objectives). Generally speaking, however, an Objective has little hope of being changed (Plasticity), then improvements are by their definition neither technically feasible nor, by extension, viable. Similarly, a lack of Transparency shatters any hope of observing demonstrations of the technical feasibility and viability of Objectives.

The concept of Desirability, appropriately for Design Thinking, has a particular focus on the end user. To that end, desirability is often considered with regards to aspects of Fit. In these discussions a useful distinction is made between Problem/Customer fit, Problem/Solution fit and User/Solution fit. We take a similar perspective, but also rationalize the stages at which each of these aspects of fit are most meaningfully assessed. Specifically, when constructing a set of candidate Objectives, delineating Fundamental from Means, and Managerial from Analytical, we are deliberately thinking about the Fit of these Objectives relative to our context: "Problem/Customer Fit". That assessment is of course predicated on the potential for identifying a Desirable solution of some kind. It emphasizes the importance of alternative, given that we are not yet in the position to confidently lay claim to "Solutions". We can't do that, clearly, until we've actually started to flesh out what actions such solutions might entail, and certainly not prior to thoughtful analysis regarding observed impact in testing. Only at those later stages (beginning with a consideration of possible Utilities and Connections in the next chapter, and into the later stages of Chapter 4) can we fully appreciate Fit as it relates to solutions.

⟡ Head-First, Belly Up

Tyler charged into the first big team assignment of the year with the full confidence of a graduate from a top engineering program. New to both the employer and industry, the hope is to prove valuable as a member of the team and someone of high career potential.

The first project assignment is to build a software system to notify customers when new products become available for purchase. Tyler designs a system that includes a model predicting which customers will be interested in different categories of new products. There's a gut feeling that success is best achieved by notifying as many customers as possible, so Tyler trains the model with the objective of achieving vast marketing reach. The drawback is that this reach comes at the expense of customer relevance.

Tyler really wants to demonstrate individuality and independent capability in this first big assignment, and focused intently on perfecting the design. Absent is a discussion, with either their manager or teammates, of the trade-offs between reach and relevance. In turn, there is no discussion of the associated assumption of priorities and design choices. When the new software system is launched, the defect is quickly apparent. The model sends marketing messages to too many customers who display low interest in the new product releases. The result is customer fatigue, and increase in complaints, and requests to opt out of future communications.

In discussing the outcomes with the manager and team, the decision is made to rebuild the model so that it strikes a better balance between reach and relevance. With additional time and effort, an improved system is created. It sends fewer marketing messages but achieves higher predictive strength in terms of engaging the right customers.

Learning Concept: **Escalated Commitment** – Sticking to, or even increasing investment in, a course of action that isn't advancing an Objective as planned.

Reflection Questions

1 What mistakes did Tyler make when initially outlining the structure of the problem, the implied objectives and implied metric for success with the new system build, and the criteria for evaluating the viability of different possible solutions?
2 What role might escalated commitment have play in the outcomes of this situation?

3 If Tyler could have a 'do over', how could they have "failed faster" and then pivoted to work with a more suitable objective?

Research Follow-up – Find out more about escalated commitment. If you end up in a situation like Tyler's someday, what can you do to avoid or at least minimize the downsides of this decision-making pitfall?

2.4 Plan Bs and Alternative Hypotheses

Among the most important points for continued emphasis in discussions relating to the selection of an FMO is that this choice is simply one amongst a set of alternatives – i.e., candidate Objectives. The reason we refer to viable objectives as "candidates", again, is because it is critical to remain open to the very real possibility that, despite looking good on paper, any elected first choice Objective might ultimately prove less fruitful than other available alternative candidate Objectives. However, this reality only becomes apparent following deep submergence into the weeds of the data, design and analysis. Having a Plan B objective (as well as a Plan C, D, E, etc.), along with well-developed alternative hypotheses bearing on how things actually work within the problem's reference system, is also very important.

Intuitively, there are numerous reasons why having a well-developed backup is useful. Having well-developed alternative objectives, for example, affords stakeholders with an avenue that their absence otherwise impedes: it grants a license to fail and, through responsible inductive reasoning, the opportunity to course correct. This benefit is available because, following time and effort spent in consideration of candidate Objectives, and working through the allotment of all but one candidate Objective onto the back-burner, swapping one objective for another becomes a simple process of replacing one of your ideas with another one of your ideas. Not entirely unappealing. Idea ownership is still your own.

Consider the alternative scenario; not having a backup. You've probably either experienced or observed the following all-too-common situation. A co-worker, a higher-level manager, a client, or even a member of your own team has been presented with evidence that coincide with either how they thought things actually were on the ground, or how they thought things would go, moving forward. For any number of reasons, the data and results just do not sync with their own mental models of the context or reduced reference system. The "should" have happened doesn't end up lining up with the "what actually did" happen. Here, alternative objectives could have helped to contextualize on-the-ground realities in a functional way – helping to make sense of divergent results. Likewise, alternative rationale could have been used to help to explain the presence of apparent paradoxes, again, offering a way to fit pieces of a seemingly mismatched puzzle together. But, in the absence of alternative objectives or rationale, when faced with conflicting evidence the interpretation that emerges is that the disconnect between their own mental model and the reality that the empirical results point to lies not with their own thinking, but with the data. Or, more often than not, unfortunately, with the analysts themselves, and their analysis.

This reaction is a textbook instance of cognitive dissonance, where in order to reconcile inconsistencies between their own thinking and empirical results, the data, or the conclusions of those responsible for generating the evidence are scapegoated as weak links in the chain. Doubling down, to minimize dissonance, colleagues, bosses or clients might even insist that a mental-model consonant course of action be advanced, despite the absence of empirical support, or even directly contrasting evidence. And, in textbook fashion, in a further escalation of commitment to a losing course of action, they are likely to get angry, hurling accusations that others are too inexperienced to know better, and push for a course of action that, ultimately, leads to a dead end.

Now, despite this broadly normative – and unfortunate – outcome, it isn't necessarily a foregone conclusion that they are always wrong. Sometime the data, the model, the engineer, or the analysis – or some combination of these drivers are, in fact, wrong. But, sometimes the individual asking the question is simply asking the wrong question or questions. In the absence of alternatives or back-up positions to help contextualize the results the team delivered, key stakeholders can feel cornered – after all, 'failures' can have broad ripple effects that echo much more widely than the results of any given project. Sometimes, failures are professionally inescapable, with long-term reputational consequences. Sometimes, they can be career ending. As these kinds of corrosive implications become more salient, these stakeholders can become more defensive, doubling down on an (otherwise avoidable) bad choice, with substantial practical repercussions with broad reach and multi-faceted consequences.

This powerful, destructive, psychologically anchored process of course can infect anyone. Entirely unwittingly, we can box ourselves into seemingly inescapable corners and self-impose extremely limiting perceptual biases and cognitive hurdles that constrain our ability to navigate within the project ecosystem in an effective or professional way. These deeply myopic experiential, psychological troughs are an almost certainty if we don't devote sufficient up-front time to the systematic development and articulation of alternatives.

In contrast, when up-front investments are made developing and building out the architecture of a range of candidate Objectives, we can expand the range of possibilities available. Taking time in advance of T1 to systematically consider a range of candidate Objectives, and then going through the process of prioritizing alternatives in accordance with the process explained in this chapter, we de facto extend out mental models of the larger systems that we work in. The availability of fleshed-out alternatives provides an operational point of departure to develop new mental models that we may have not thought about sufficiently prior to that point. In addition to expanding available degrees of freedom, we are also in position to appreciate Connections between candidate Objectives, hierarchies among Mean Objectives and Fundamental Objectives, and tradeoffs and synergies among them. Importantly, and this is worth emphasizing, we also, attain operational and psychological ownership of these ideas. This inspires us to take the high road in closing the gap between the "should" and the "did". Pulling those other great ideas off the backburner, and honestly confronting the possible shakiness of prior assumptions. We allow ourselves to evolve more, and block less. We embrace faster failure, and better learning.

Let's be clear here. The savvy leveraging of alternatives is NOT the same as giving up. It is not the same as satisficing. What it is keenly reading the analytic and practice nature of the room. It's taking a turn to avoid a crash. It's being smart.

The value of alternatives also applies to other stages of The Cycle, as we discuss next.

Practitioner Recap

At their essence, Objectives describe the reason for embarking on a project and the means by which success, or failure, will be determined. Teams will need to get comfortable with the fact they are driving toward precision in what to accomplish, but in having to use an imperfect measure, as they will learn more along the way. Mapping objectives and communicating them in a way that is relevant to the recipient can reveal progress, increase understanding of work complexity, and increase the likelihood that stakeholders stay engaged and supportive of the challenging work required to achieve Managerial Objectives of any significance. Fundamental Objectives describe what is to be accomplished; Means Objectives describe how it will be accomplished; Managerial Objectives describe the motivation of the work; and Analytical Objectives describe measures of success in the techniques that deliver the Managerial Objectives. Creating, and continuously evaluating alternative objectives will help to ensure that teams do not become anchored on Objectives that may become irrelevant as the work progresses and they learn more. It also helps to ensure that teams are mentally prepared to pivot as what they learn ultimately challenges earlier assumptions.

Note

1   All quoted definitional terms are drawn directly from page 20, Lewrick, Link, and Leifer (2020).

Chapter 3

The Shape of Causes and Correlates

It can be tempting to attempt to jump directly from a stated Objective, managerial or analytical, to a solution that our gut tells us might advance it. Easy, available solutions to well-thought-out questions offer convenience and timeliness. They can be of high value. But they can also be of high risk when their implications are not considered sufficiently and systematically to at least some degree. We are reminded of the risks of insufficiently considered solutions in classic parables such as the following.

> *Temptation had its way with my companions*
> *And they untied the bag.*
> *Then every wind*
> *Roared into hurricane the ships went pitching*
> *West with many cries; our land was lost.*
> — R. Fitzgerald's translation of The Odyssey (p. 166)

The above translation is drawn from the prologue of Book Ten of *The Odyssey*. In this narrative, Odysseus, having been granted smooth sailing by the keeper of the winds, Aiolos, finally sees his homeland in sight. Exhausted from manning the sails for the last nine days straight, he goes to sleep, leaving the 'last mile' to his crew. Unfortunately, he never did tell them what was in the large bag that Aiolos had given him, tied with a silver cord ...! The crew, being (unfortunately and tragically) curious about any riches the harrowing journey might have yielded, felt that this might be their last chance get a fair share of the spoils picked up along the way. Ultimately, their efforts to uncover the hidden details contained in Aiolos' bag cost all of them their lives. And, for Odysseus, their fatal curiosity cost him many more years at sea.

This vignette offers a beautiful example of how classic, in many ways, omissions of details can utterly derail the otherwise promising trajectory of a project well on its way to success – or at least perceived to be heading

DOI: 10.4324/9781003427650-5

there. Often, the crew gets the brunt of the blame for what happens in this anecdote. From a broader vantage point, however, Odysseus who typically regarded as the cleverest of Greeks in the Homeric epic, can't escape criticism. Ultimately, it turns out that he'd made a bush-league blunder in team leadership and transparency here. Homer offers no rationale for why Odysseus didn't simply tell his crew exactly what was in Aiolos's bag, perhaps offering them something along the lines of: "This is a bag of crazy powerful wind. I have no idea why we have been made carry it with us, rather than leaving it back on that island. It seems to be a pretty obvious test of self-control, though. You know the Gods ...! If we open it, we are probably going to be blown off course, so ... just don't!... Napping now. Wake me up when we're there."

There are lots of reasons individuals choose to not share information with coworkers. Sometimes, information is withheld because of perceived advantages associated with being an information gatekeeper. This framing reflects the belief that having unique access to information is likely to be 'useful' at some point in the future. Sometimes, otherwise important or relevant information is withheld from others because of anticipated reactions to the information being shared. This idea is reflected in often-repeated caution 'don't blame the messenger', when unappealing or jarring details are shared. Information can also be withheld because people may be afraid of being seen as foolish if they think there is a reasonable chance the details being shared are already widely known or obvious; or if they suspect the details either aren't entirely accurate or well thought out. In short, information may be withheld because individuals feel insufficient psychology safety, and refrain from conveying specific information because of the perceived risks associated with doing so – a concept we return to below.

At the same time, whether sailing the last mile of an epic voyage or working through a forensic process decoding why key links in a supply chain broke down, there are real costs associated with withholding-back essential information and details. This reality applies to details and information bearing on factual characterizations of what is known (e.g., 'the bag contains wind'), to the details of one's mental models of cause-and-effect (e.g., 'if you open the bag, we are going to be thrown off course'). This is why it is generally critical to have an accessible way to elucidate the details of (1) what is known, (2) what is assumed, and (3) what is discovered as engineering and analytics teams embark upon and progress in their journey toward value and impact.

In our discussion of Objectives, we explain the central role played by thoughtful consideration as a way to advance understanding of what we might accomplish. In the following discussion of available levers to pursue those Objectives, we draw on aspects of that discussion, applying a similar set of perspectives and focus on details, and also explore key aspects of the potentially complex nature of their impact.

3.1 Utilities: Levers to Impact and Advance Objectives

As part of the exercise in Chapter 2, you spent time considering Objectives associated with a recently completed project. As you recounted actual Objectives in place at that time (as well as alternate candidate Objectives), you rated the value of each of these Objectives, which ultimately included an evaluation of their transparency, plasticity, and fit.

Returning to his exercise, select one or two of these Objectives that seemed to rise to the top in this multi-faceted evaluation. It might be that these were the same Objectives made obvious in the actual project, or they may be others you identified retrospectively. Answer the following question with these Objectives in mind.

Question: What kinds of things do you think you have (or had at the time) control over that could impact these Objectives in the desired direction?

These 'things' could include anything from budget allocations, to machine time, to skilled personnel. It is useful to get specific here. Perhaps key stakeholders had funds available from the budget to purchase access to data? or invest into training? or hire crucial personnel? or requisition new equipment? Perhaps influential project personnel managed a number of direct reports, and had the discretion to allocate their time to specific work? Or had discretion over how specific pieces of equipment were used? Perhaps there were quality threshold considerations that could be modified? Or service guarantees that were subject to determination or adjustment. Maybe there were design or marketing parameters that would need to be specified, some of which had direct bearing on the focal project outcome?. Whatever these 'things' were, if you felt that you, or members of your group had some chance of influencing these, with the associated potential of using them to advance your Objective, jot these down in Table 3.1.

In follow-up, do you believe that any of these 'things' would remain relevant if the focus of the project was changed from one candidate Objective to another? Place a check mark next to each of the items in Table 3.1 that you think could be utilized to advance Objectives other than those Objectives you've been thinking about.

You may have noticed the emphasis on the term 'utilize' here. This is deliberate. We are looking for 'things' that are useful here. Not simply in the sense that they might predict *any* Objective outcome, but rather in the sense that they might be controlled with the intent of achieving *a particular* Objective outcome. It is the availability of these 'things' for deliberate

Table 3.1 Things You Can Utilize Towards the Advancement of Objectives

i.	
ii.	
iii.	
iv.	
v.	

manipulation, with the goal of advancing Objectives, that make them what we refer to as Utilities, rather than simply 'variables'. More formally, here we offer a definition of Utilities:

> **Utilities**, in the OUtCoMES Cycle framework, describe options that decision-makers have control over (either direct or indirect control), and which through that control have the potential to advance specific Objectives. They can take the form of decision variables to be optimized, or coefficients in predictive models to be estimated. Like Objectives, the conceptual identification of Utilities draws on real-world considerations, while their numerical specification benefits greatly from analysis.

Rather than being purely mathematical, Utilities have a distinctly Objective-oriented, practical designation. If you can't imagine influencing something, there is little point in thinking of it as a lever with potential to impact an Objective. If isn't useful, it's not a Utility by this definition. At the same time, if you don't have any evidence describing the relationship between something and the focal Objective, you will have real difficulty advocating for its use, even if you control it.

However, if given the opportunity (coupled with willingness to take responsibility!), access to relevant data and resources affords wide-ranging control over where a project can go. For example, although stakeholders ultimately have control over the estimation of predictive models, it is often allowed that full control of prediction is in the hands of algorithms and software packages (i.e., we occasionally encounter that claim as a bit of a crutch). Although, in reality, stakeholders have considerable discretion over what models are run, what those models include as predictors, and how those predictors are structured. As a result, a great deal of control is retained over estimation of effects, even if indirectly. The hope here, of course, is that such control is used so that predictions are robust against a wide range of assumptions.

The same applies to prescriptive efforts. Subject to constraints specified by key stakeholders, the focus is on features of the reference system that can be changed to determine the best possible approach to advance a particular Objective. The choice of Utilities (i.e., decision variables in the language of optimization) is of particular relevance because of the expectation that these can be changed; and also because they are believed to be of relevance to the Objective. It is understood (or believed) that either direct or indirect influence over Utilities matters to the Objective. Things that can't be influenced are either treated as constants or, if they don't relate sufficiently to the reference system, are simply not included in the analytical model.

Control and consequence. Influence and relevance. These are key to identifying what can (and should) be done in pursuit of Objectives. They are the core attributes of effective Utilities.

Nevertheless, identifying what could, or should, be the focus of this process isn't always straightforward. The search for candidate Utilities, these consequential and controllable levers, starts in much in the same way as consideration of candidate Objectives. Targeted questions are asked to determine analytical viability, with the goal of establishing relevance and potential influence. Candidate Objectives provide critical context for evaluating and prioritizing the most relevant Utilities. After reading Chapter 2, these questions will look very familiar.

Questions to establish Utility transparency:

- In what unit is the Utility is measured? (dollars? days? A count? A percentage?)
- What is the current level of the Utility, and/or what is a typical historical measure of the Utility? (i.e., the 'current state', or current state average or median).
- How often is data collected on the Utility?
- By whom and under what circumstances are data collected?

Think about predictive analysis. If you don't have a good under-standing of (or access to) data relating to a given predictor, there's little chance of effectively estimating its impact on what you're trying to predict. In a prescriptive sense, if the range of values a decision can take on isn't known, it's extremely difficult (if not impossible!) to establish effective constraints for that decision. As with Objectives, Utilities can be scored based on their transparency (i.e., 1–5, with 1 = Very Poor, 5 = Very High). Utilities with low transparency might be consequential to a given Objective (i.e., in a Platonic sense), but it is difficult to imagine controlling these effectively absent evidence bearing on those consequences.

Questions to establish Utility plasticity:

• Is there variance in data describing the Utility which suggests the Utility can reach more desirable levels? Is the Utility already at, or close to, a limit?

Even if a given Utility is highly transparency, much as with Objectives, control will tend to vary across Utilities. Some levers are simply very difficult to push. Some are, essentially, stuck; perhaps due to recent organizational incidents or strategic mandates that have left them frozen. Or, perhaps they have simply hit their limit in the direction that would otherwise advance an Objective. Others could just be a bit rusty – 'We've pushed on them in the past, but it's been a long time, and we might need some reminders as to how.' Analogies aside, it is essential to rationalize the degree of control across all candidate Utilities. Applications in prescriptive analysis, and in design, are perhaps a bit more obvious than in predictive analysis. It might be hard to image not being able to estimate the effect of something on an outcome. However, if that something has seen little change over a considerable period, use in near-term prediction may be limited. Once again, rating Utilities in terms of their likelihood of yielding control is helpful for establishing priorities (1–5, with 1 = Unlikely to be moveable, 5 = Very likely to be moveable).

Questions to establish Utility fit:

• Does a change in the Utility align well with what is going on now? Is the Utility something key stakeholders are in the position to influence effectively (even assuming the best data available)?

If the predicted and prescribed applications of Utilities run counter to the organization's dominant paradigm, even if Objectives are well aligned, engineers and analysts can face some real hurdles and pushback. Again, organizations maintain a wide range of Objectives, and seek to advance many of them simultaneously. Even if any two Objectives might be 'viewed' as independently pursuable, the adjustments that might most readily advance one Objective might work counter to the advancement of the other, either directly or indirectly. Thus, understanding the broader context can be crucial for ensuring the ultimate applicability of the intelligence which project teams develop. As a consequence, it is essential to evaluate candidate Utilities on the basis of fit as well (1–5, with 1 = Poor Fit, 5 = Strong Fit to broader organizational interests).

The best Utilities are going to score high on these three criteria, as are the best Objectives they target. If strong candidate Utilities can't be identified ... there are only a few options available: (a) integrate the input of other key stakeholders, and reassess available options to identify additional candidate Utilities, (b) return to the initial list of candidate Objectives, and determine if an alternate candidate Objective might have a broader range of strong Utilities associated with it. It is extremely important to be flexible, and recognize that it may sometimes be necessary to cycle back. It is far better to take a few steps backward in order to find the right path than to push forward in a direction that leads over a cliff.

If a strong Objective, or short list of these, can be identified, and then also paired with controllable and consequential Utilities, we can start to break out some of the finer details on how to connect these dots.

🌐 *A Scaling Snafu*

Maurice is asked by his team leader to handle capacity planning for critical hardware resources. These particular resources will be needed during the upcoming holiday shopping season as they are used to power various types of automated marketing programs. He starts to develop a scaling function for each of them to help provide projections for the levels of hardware capacity required to deliver all marketing campaigns during this busy period.

After analyzing an initial subset of his team's marketing programs, Maurice sees what he feels is a common pattern in their scaling functions. It shows that the need for hardware is directly proportional to the overall increase in customer traffic on the website's home page. Extrapolating from this small-sample observation, he projects the hardware capacity needed for all marketing programs and finishes his task ahead of schedule. Maurice then gets finance approval to order the additional cloud capacity needed for the holiday season.

One month into the holiday season, Maurice's team leader stops by his office looking distraught. A large queue of backlogged marketing campaigns has built-up. It's having a material impact on the company's financials. Leadership demands immediate answers and an explanation of where the problem lies. Maurice opens an investigation, reviews the team's capacity dashboard, and sees that all available hardware has been at maximum utilization for over a week now. Maurice comes to the realization that he ordered additional hardware based on some incorrect underlying assumptions.

After digging further, he determines that a third of the marketing systems that he projected capacity for are not scaling linearly with the overall customer traffic to the homepage. The level of customer engagement with the website's shopping cart feature appears to be a more consistent scaling predictor. Maurice observes that the conversion rate of home page traffic to shopping cart traffic is significantly higher now during the holiday period than it is during the off-peak season. This differential led him to significantly under project the needed hardware capacity for those marketing programs.

After sharing these findings and consulting with his team, Maurice reevaluates the scaling functions for each distinct marketing program to ensure that he has identified all of the correct inputs. He uses this new data to recompute the team's hardware needs, placing an additional cloud capacity order which, once available, is able to drain the queue of backlogged marketing campaigns and to get the team's programs back on track for the holidays. The team leader reiterates that the delay in marketing campaigns has been costly, and that the scaling function for each individual marketing program must be well-documented with peer review for accuracy, moving forward.

Learning Concept: **Activity Trap** – Mistakenly associating choices (e.g., in Utility identification), and related activity, with progress towards an Objective. The trap can often be accompanied by a lack of post-hoc effort to monitor evidence of presumed, and alternate, causes and their effects.

Reflection Questions

1   To what extent does the ability to achieve Analytical Objectives, and advance FMOs, depend on the identification of proper levers (Utilities)?
2   How can you explain Maurice's susceptibility to an activity trap in this case? Was it insecurity, overconfidence, time pressures, or something else that caused his failure to identify and utilize superior Utilities?
3   Did the team leader handle the situation correctly? Did Maurice get off too easily after making this error in making his scaling functions?

Research Follow-Up – See what you can find about the activity trap in decision-making. What can you discover that might help you avoid future errors in high-stakes situations?

3.2 Connections: The Nature and Course of Impact

Much of the analysis underlying identification of effective Utilities tends to be either categorical/binary (i.e., 'yes this might be useful') or ordinal (i.e., 'this is likely to be more useful than that'). But the actual relationships between Utilities and Objectives, the dynamics of cause and effect, often aren't that simple. Objectives and Utilities can be described using any number of designations, from categorical, to ordinal, to interval (i.e., equally spaced values, often on a continuous numerical scale). Their connection, or at least how they appear to relate to one another, can therefore be described using any combination of these various measures. Imagine FMOs along the lines of 'we want to better-anticipate product failures' or 'we want to reduce product failure rate', with associated Analytical Objectives like 'increase the maximum likelihood of failure prediction' or 'further minimize current failure rate (i.e., achieve $x\%$; get closer to 0 than our current state)'. While the kinds of Utilities available might be categorical (i.e., 'use this machine or technique rather than that one'), ordinal (i.e., 'do these process steps in this new sequence'), or interval (i.e., 'invest this much money into better inputs'), the outcome being predicted or advanced is a percentile or binary outcome, depending on the focus of the analysis (i.e., a single product or a batch).

The complexity of the relationship raises a number of questions bearing on how we think about and analyze the connections between the underlying aim(s) of the process (i.e., Objectives) and the levers available to accomplish these aims (i.e., Utilities). As with these first two components of The Cycle, it is useful to have a working definition of Connections as we start to take a closer look at how these elements of the Cycle all fit together.

Connections, in the OUtCoMES Cycle framework, refer to relationships between and among Objectives and Utilities that capture cause-and-effect or coincident association with other factors, such as limits (constraints). Connections can have deterministic and mechanistic characteristics (anticipated by definition), as well as stochastic and seemingly random characteristics that rely on prediction and understood distributions in risk. Connections may involve seemingly simultaneous or lagged relationships, as well as feedback mechanisms involving subsequent reinforcement phenomena, or counterbalancing phenomena among Objectives and Utilities.

There's definitely a lot to this definition. We will address each of these elements in turn, since many of these elements are far too overlooked in practice. Ultimately, the key to effectively understanding Connections is to start with what you know, or at least with what you can discern through some systematic sensemaking. This serves as the foundation for initial mental models that are subsequently adjusted as evidence from analysis emerges.

To be clear, initial thoughts regarding Connections are often reflect past experiences and observations. However, they are also often tantamount to hypotheses, or at least propositions, regarding 'what might' or 'what is expected' to play out in the future. Referring back to our earlier discussion of design thinking concepts of Fit (Chapter 2), it is at precisely at this stage that semblances of "Problem/Solution Fit" and "User/Solution Fit" begin to meaningfully take shape. However, the ability to make these assessments won't full materialize until we construct and test models that permit the observation of explicit solutions. That is, organizing the pieces isn't equivalent to having a fully testable model (Manifest) capable of surfacing solution that fit problems (Explicate), which in turn can be Scrutinized for fit and thus more comprehensive assessments of "Desirability". Nevertheless, just as with our efforts to brainstorm around Objectives and Utilities, brainstorming here is a critical step in helping us move down this path. Design thinking activities such as the 5W+H can go a long way in this regard (Lewrick, Link and Leifer 2020, p. 68), in that repeated questioning of the paths between Utilities and FMOs, i.e., Connections, further shed light on the role of other possible Utilities and Connections.

Throughout the process of developing a list of relevant Connections, it is also important to look for evidence through a wide spectrum of lenses. It is entirely too easy to develop an inaccurate picture of a landscape when only a single snapshot is taken. To demonstrate this, let's go through a little exercise.

🚴 *Exercise:* For this exercise, you'll first spend some time thinking about data, and the kinds of insights it is possible to extract from limited visual depictions. Visual depictions are a critical component of any portfolio. But, they also are fundamentally limited because any given depiction can reflect on only a limit set of aspects of the key Connections in focus. In Figure 3.1, for example, there are four data sets. Each set of data has an x and a y variable. Typically, the definitions of these variables, their units, and the context would all matter. However, for this exercise this information about the data sets isn't provided! Regardless ..., for each graph in Figure 3.1., draw a line that you think captures the relationship between x and y. Each graph can have a very different line shape, so take a minute to reconnoiter and then give it a shot.

Now that you've had a chance to lay out what you think is going on between these variables running blind, take a look at some of the descriprive and predictive statistics associated with each of these data sets. You might be surprised to learn that the means and variances of x were the same in all four data sets (9 and 10, respectively).

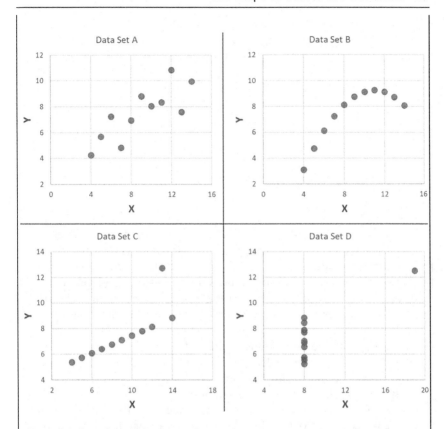

Figure 3.1 Four Sets of Data.

The same holds for *y*, with a mean of 7.5, and a variance of 3.75 in all four sets. You might be even more surprised to learn that a 'best fit' straight line, estimated with any basic statistical package running standard linear regression, will have exactly the same form in all four cases. The best-fit lines and fit equations for each of these cases, along with the fit statistic ($R^2 = 0.67$ in all cases), are presented in Figure 3.2.

Is something wrong here? Are your eyes deceiving you, or is the analysis deceiving you? In the spirit of full disclosure, this is a special quartet of data sets, based on what is known as Anscomb's Quartet. They were created in 1973 by the statistician Francis Anscombe. Their intent was to show that (limited) statistical summaries alone can be prove to be insufficient for providing a clear picture of the nature of what we refer to here as Connections. We could also say that looking at a single visual depiction can be very misleading. Ultimately, understanding the multi-faceted nature of real-world phenomena requires application of multiple tactics.

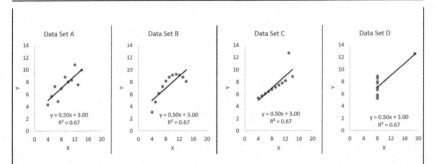

Figure 3.2 Best Fit Lines for Four Data Sets.

In this example, there's nothing fundamentally wrong with the mechanics of the best-fit analysis. But, we also aren't being forced to model the relationships as straight-lines in the regression; that is only being done to illustrate the problems associated with relying on a single perspective. If there is reason to expect non-linearity or discontinuity, we need to pull on that thread rather than disregard it. Of course, that thread might lead nowhere. We might find that what we saw in the data – or thought we saw – was simply a mirage. But it might also be something significant. If we don't "look" at the evidence through multiple lenses, we risk a great deal; misdiagnosing, overfitting, and generating poor prescriptions, all because we didn't push the boundaries of our understanding.

Consider the classic parable of the blind men and the elephant, that can be traced as far back as the fifth-century BCE Buddhist *Tittha Sutta* text. Each of these men uses a singular perspective to diagnose the exact nature of this creature, which they've never encountered before. In turn, each comes up with a very different assessment of the beast (i.e., a snake, a tree, a wall, a rope) and, as one might imagine a distinct prescription for what to do with it. Yet, ultimately none of these would prove effective. Only a far more coordinated and systematic examination of the elephant would allow these men to come close to understanding what they actually have on their hands.

In order to meaningfully move forward in leveraging Connections between Utilities and Objectives, it is critical to scrutinize the nature of these Connections in a purposeful way.

This isn't a trivial caution. It is certain that pitfalls will emerge in the process of describing Connections. Returning to the definition of Connections, some are bound to be somewhat straightforward. Sometimes, there is a clear formulation that translates changes in a Utility to an unambiguous shift in the level of an Objective. For example, each volumetric decrease in a specific material, as part of an overall design, may consistently impact the total weight of a product (i.e., a weight targeted for reduction as part of a Fundamental or Means Objective). Or, the amount of money spent on training (i.e., a possible Utility), is limited by, and contributes to the use of a total budget (i.e., which may be fixed, or may be targeted for reduction). Similar examples can be drawn from manufacturing resource planning contexts. For example, in simplified form, assume that it takes N working components, M units of energy, and L worker-hours to produce a single machine assembly. What inputs would one anticipate needing if 70 machines were demanded? 70xN components, 70xM units of energy, 70xL worker-hours. Some Connections largely emerge from the direct description and application of physical laws and institutional rules. The 'math' is clear, and often fairly simple. Prediction is perfunctory.

However, that clearly isn't always going to be the case, both because there is uncertainty about the true nature of a set of Connections, and/or because there are factors that can't be controlled influencing how causes yield effects. Anyone with experience in this kind of setting knows that what we'd like to happen doesn't always (or even often!) happen. Workers aren't homogenous automatons, and thus there bound to be differences in productivity and manufacturing accuracy across shifts and work units. Individual workers also are likely to vary in their outputs both within and across time. It's also easy to make the conservative assumption that the components acquired for assembly also are likely to vary in quality, at least to some degree as is typical in the real world. We could model this variability by assuming that out of P parts procured, on average only Q are of sufficient quality to be deemed useable. More specifically, the actual number of useable components (n) from any order of P components tends to resemble a quantity drawn from a Normal distribution, with a mean of Q, and a standard deviation of R.

While it may be largely possible to describe this variation, the best we can ever hope to do is plan for it. Here, one of the Utilities in this model is how many units to order, with the intention of obtaining N useable components (i.e., or some multiple thereof depending on need). We can think of the chances of getting *at least* the number of components we want as a Means Objective, working toward the Fundamental Objective of getting a specific number of machine assemblies made (i.e., 70 from the example above). In this case, it is likely going to be necessary to order more components than are actually needed to generate these 70 assemblies, given the evident risk. We accept that the Connection between how much we order (i.e., a Utility) and how many machines we can make (i.e., an Objective), in conjunction with the costs associated with being short of units, or having excess components (i.e., perhaps Connected to other Objectives, or limiting constraints related to budgeting or storage capacity), is something we can only estimate. Although prediction is critical in defining Connections, its accuracy is not a foregone conclusion.

Risk is a reality. It exists. Whether it is appropriately accounted for, or planned for, risk plays a significant role in how understanding of the systems we are working in develops, how we put available Utilities to use, and how we work towards FMOs.

As reflected in the definition of Connections, Risk is one of several complications defining the emergence of predictive and prescriptive models. Unfortunately, although Risk plays a fundamental role in relationships between Utilities and Objectives it is, however, all too often underrepresented in analysis. The other three features of Connections, which are presented in Figure 3.3 are Constraints (i.e., which include limits to the appropriateness of linearity assumptions), Lags in time between cause and effect, and Feedback mechanisms. We discuss each of these dynamics, in turn, below.

Constraints take many forms in the real world, and play a key role in engineering and analytics work. As has already been noted, constraints may apply both to Objectives and Utilities in a univariate sense. Most things simply have bounds that cannot be surpassed. A glass can't be filled beyond 100% of its capacity. Once it is full, it's full. Continuing to pour water into a glass won't change this constraint. Similarly, when the glass is empty, no matter how much effort is exerted, it isn't possible to surpass the limit of 0% capacity utilization. The same kinds of constraints apply to a wide variety of things we might want to manage or have an impact on. While it may seem obvious, bounds have significant implications for both predictive and prescriptive analysis. Ignoring bounds is likely to yield implausible analytical results and prescriptions

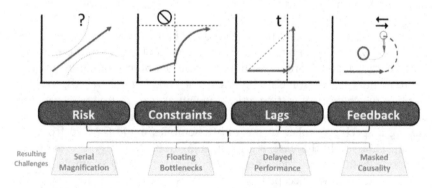

Figure 3.3 Four Features of Connections Often Underrepresented in Analysis.

that simply don't make any sense. In contrast, meaningful recognition of bounds requires explicit consideration of the methods applied. For example, in the predictive analysis of categorical or count-based measures, methods distinct from those used to predict Normally distributed continuous outcomes are used. In prescriptive analysis, bounds must be specifically described to ensure that Utilities and Objectives don't take on values outside of real-world ranges.

However, there are other implications typically associated with bounds. Some Objectives and Utilities are far harder to manipulate as they approach bounds. Think about the glass of water analogy. While filling the last 1% may be no harder than first 1%, emptying the last 0.1% can be particularly difficult. Hydrogen in water likes to bond, which includes adherence to the interior surfaces of containers. In a more industrial sense, perhaps, consider what is needed to implement a six-sigma level of quality (3.4 errors per million), in contrast to three sigma (66.8 thousand errors per million). As we strive to squeeze out the occurrence of errors, approaching the theoretical limit of 0, the investments required to do so, in personnel, machine maintenance, material inputs, ambient control, etc., can grow at far more than linear rates. If non-linear dynamics in the per unit cost of these asymptotic improvements are accounted for, the reality of the challenges to such improvement can be greatly misrepresented.

Likewise, insufficient appreciation of possible lags and feedback mechanisms can make the products of analysis, and hence those producing it, subject to considerable criticism. If past modifications to operating processes took several days, or even months, to yield returns, appropriate consideration of such timeframes must take place in both predictive and prescriptive efforts. If actions involving the use of certain resources tend to translate into a lack of their availability in the future, or if they might inspire actions by others that run counter to FMOs, ignoring such dynamics can also leave supervisors and clients asking questions like 'why did we hire these people?'. The existence of simple lags can be evident through contextual experience as well as through analysis. Feedback mechanisms can be substantially more difficult to identify, but certainly benefit from having multiple stakeholders involved in the discussion of Objectives, Utilities, and Connections. Or at least ensuring that implicit mental models, formed and held by project teams, are made visible to these stakeholders. For those dealing with complex systems with clear feedback mechanisms in place, one nice primer to help start thinking about the nature of the stocks and flows inherent to these would be 'Thinking in Systems' (Meadows 2008).

To recap, Connections take many forms in bridging Utilities that we might influence in our pursuit of FMOs. In fact, the "Co" that represents Connections in The OUtCoMES Cycle, can help us remember that.

Connections	
Risk	Contingencies and *(lack of)* Control
Constraints	Containment & Conditions
Lags	*(lack of)* Contemporaneousness
Feedback	Co-dependencies
Tools:	Counter arguments, Concept Maps

Figure 3.4 Phenomena entailed by Connections between Utilities and Objectives.

Connections can certainly describe basic *co*rrelations, ideally causal, that can exist between and among Utilities and Objectives. They can also, however, describe *co*ntingencies that are out of our *co*ntrol, and which may give rise to risk in these relationships. Connections may entail constraints that impose *co*ntainment on the discretion we have over changes to Utilities or Objective, or detail *co*nditions limiting relationships. They may lack often-desired *co*ntemporaneousness, or sufficient temporal proximity, between cause and effect. They man even involve *co*-dependencies wherein changes imparted on Objectives by way of Utilities subsequently push back or reinforce changes in those very levers. Challenging your intuition regarding connections via *co*unter-argument efforts, and visualizing connecting tied via *co*ncept mapping (a form of which we will present in the next section), can be particularly helpful supplemental brainstorming tactics for identifying such phenomena where relevant. For general reference in both initial brainstorming efforts as well as later evidence-driven revision, in Figure 3.4 we summarize these various possible characteristics of Connections.

3.3 The Virtues of Structured Documentation

With this last point in focus, it is worth reviewing some of the more meaningful features of Objectives, Utilities, and Connections, and how the documentation of these can be used to increase the likelihood of generating strong project returns. The benefits of effective documentation emerge not only from the ability to share ideas beyond the boundaries of the project team, but also, and crucially, as a living history of our key ideas, revealed evidence, and evolving alternatives that emerge during a project's course.

We prime this focus with consideration of key attributes of Objectives identified in the discussion from the previous chapter. There, Managerial and Analytical Objectives were distinguished from one another. Analytical

Objectives were identified as a mechanism for codification of evidence and quantifiable guidance in pursuit of the real-world improvements implied by Managerial Objectives. We also crossed this categorization with an explicit distinction between Fundamental and Means Objectives. This was done in recognition that end-goal FOs often rely on the accomplishment of multiple discernable, and ultimately pivotal, intermediate steps. In our discussion of which Objectives were most likely to yield both evidence and actionable intelligence we also, as in this chapter, discussed several additional points with potential to mitigate our Objective foci. These points include the extent to which Objectives are Transparent, their apparent Plasticity, and Fit. Associated considerations lead to inclusion of the current state of Objectives, and any relevant natural limits (i.e., bounds or univariate constraints). We noted the value of understanding the proximity of Objectives to these bounds, and addressed the issue of proximity in the current chapter as well.

At this point, we can begin to summarize some of the Objectives considered in Chapter 2 by way of tabular record-keeping. Table 3.2 provides an example table into which these details might be structured and organized.

In the first column of Table 3.2. you can provide labels or short descriptions of the focal Objectives. In the second column, you can specify the Objectives as being either Fundamental or Means Objectives (mark either Fnd (for Fundamental Objectives) or Mns (for Means Objectives). Subsequent columns can be used to document scaled assessments of Transparency, Plasticity, and Fit (e.g., on the 5-point scale described earlier), an understanding of the current state of measurement, and associated bounds. While the intention is to ensure some form of mapping between Managerial and Analytical Objectives, because of key distinctions in their core purposes, we account for each type of Objective separately (i.e., Analytical Objectives below Managerial, symbolically reflecting their supporting role).

At this point two critical points are worthy of note. First, the intention here is to allow ideas to be documented. Not only is it important to flesh

Table 3.2 Example Structure for Objective Documentation and Tracking

Managerial Objectives	(Fnd / Mns)	Trns	Plst	Fit	Current State (w/ units)	Bounds
Analytical Objectives	(Fnd / Mns)	Trns	Plst	Fit	Current State (w/ units)	Bounds

out the Objectives that you 'think' are the most important to your current view of what a project can accomplish, it is also absolutely critical to document your second and third most likely candidate Objectives as well. This admonition applies to both Managerial and Analytical Objectives. Rows can be added to the table as necessary depending on the number of candidate Objectives identified. Some of the highest-returning projects begin with two, three or more alternate Managerial Objectives (i.e., Plans B through D). Some might begin simply as Means Objectives, but ultimately emerge as the highest value pursuits in the set.

This leads us to the second critical point – expect the content of this table to change. Allow it to change. But, keep prior versions of the table. It's useful – essential – to look ahead, but it can also be extremely valuable to look back on prior assumptions, earlier assessments, evidence that led to dead ends, and the origin and inspiration of new ideas. Forcing yourself to document multiple alternative Objectives encourages intelligent pivoting as we confront the scrutiny that analysis ultimately provides.

There is virtue in forcing ourselves to structure our best ideas, as well as in forcing ourselves to consider ideas that present alternative pathways for value.

Similar value emerges from efforts to document and track understanding of Utilities. Consider the summary structure presented in Table 3.3. Here, since there are often a far greater number of Utilities than Objectives, we consolidate some of the rating fields (T/P/F provides locations for transparency, plasticity, and fit accounting), and combine the fields for current state, and bounds (i.e., since ultimately one is a subset of the other).

Since no two projects are identical, it should be emphasized here that the specific structure reflected in Tables 3.2 and 3.3 are only examples of how this process of accounting can be done. Rows and fields can be added or eliminated, consolidated or expanded up as suites the needs of the context. It is possible to imagine additional fields (columns in the table) or designations to specify whether specific Utilities are unique to predictive rather than prescriptive efforts, for example. Or whether certain Utilities focus on a particular Means Objective, as opposed to being identified as directly influencing an FMO. The importance of one structure over another,

Table 3.3 Example Structure for Utility Documentation and Tracking

Utilities	T/P/F	State (& Bounds)	Utilities	T/P/F	State (& Bounds)
	/ /			/ /	
	/ /			/ /	
	/ /			/ /	
	/ /			/ /	
	/ /			/ /	

Table 3.4 Example Structure for Connection Documentation and Tracking

Connections (and Utilities / Objectives involved)	Current State (& Slack)

or one designation over another, should be allowed to emerge over time, just as evidence and a sense of the true nature of the reference system must be embraced in identifying project opportunities and high-value foci.

A similar accounting approach can be leveraged to evaluate the value of Connections in the systems we work to understand and improve. One example of a documentation structure for tracking known and presumed relational dynamics between and among Utilities and Objectives is presented in Table 3.4.

In this instance, and for contrast, we provide a much more narrative form of structure, given that the complexities of Connections are not easily distinguished by a handful of quantities. Outside of a codified model (i.e., which we address in the following chapter), Connections may be more effectively articulated using descriptive references to the Utilities and Objectives that they impact. Univariate bounds are already captured in the documentation structures tracking Objectives and Utilities. The Connections described in structures such as the template depicted in Table 3.4 tend to focus on multi-variate relationships (e.g., the impact of one Utility on an Objective is amplified by the level of another Utility), and limiting constraints (e.g., three Utilities relating to spending can't exceed budgetary limits).

In light of the fact that each of these documentation structures is part of a larger story, and really only represent the foundation on which more systematic modeling and estimation rests, it's also useful to step back and see how they can be seamlessly integrated into that whole.

As depicted in Table 3.4, each of these structures, and the discussion offered to this point, take us just halfway through The OUtCoMES Cycle – the O (Objectives), Ut (Utilities), and Co (Connections) as described in the overview presented in Chapter 1. Now that we have considered some of the minutia associated with these foundational elements of the Cycle and, hopefully, have begun to come to terms with the fact that their identification begins with a great deal of guess work (i.e., and thus is ultimately subject to reconsideration through the Cycle), we are ready to start putting these pieces together. The broader documentation structure (i.e., previewed in Chapter 1), that encompasses details such as those included in Tables 3.2, 3.3, and 3.4, is the Systems-oriented A3 (or S-A3) framework. Figure 3.5 provides a close-up of only the left side of this framework. The other side of the framework is built out in Chapter 4.

As is apparent from the S-A3 rendering in Figure 3.4, additional space is deliberately provided to include descriptions of the origin of the current problem or project focus, the detailed nature of the current state of the

Theme: **Owner:**

Background and Problem:
{Insight how the current (aka, pre-analysis) context developed, and existing trajectories. Clear and tangible depictions, leveraging effective visual depictions of critical processes and outcomes are particularly valuable here}
{Insight how the problem is perceived, including what performance improvements are desired; providing input into candidate **Objectives** for further consideration. Such Objectives are distinct from, but a step towards prescriptive "goals". Here it is critical to outline BOTH **Managerial Objectives** (e.g., ROI to be maximized, turnover to be minimized, recruitment increase, loss reduction, market capture growth, etc.), as well as Analytical Objectives, if distinct (e.g., R^2 increase, accuracy maximization, Type-II error reduction, maximizing the ration of between to within group variance, etc.). The table below can be used to track these ideas (you may have more of one than another). NOTE: IF you are doing a substantial amount of Predictive Analytical work, or exploratory Descriptive work, regardless of whether you are trying to model a continuous variable or a classification, you must outline some metrics useful in gauging the performance of your analysis; Both in the managerial and the analytical sense, we have not hope of improving things if we can't measure how good they are and how good they can get}.}

Managerial Objectives	(Fnd / Mns)	Trns	Plst	Fit	Current State (w/ units)	Bounds

Analytical Objectives	(Fnd / Mns)	Trns	Plst	Fit	Current State (w/ units)	Bounds

Current State Details:
{Insight into nature of above bounds, priorities, and any inter-relatedness/feedback among above. This should inform descriptive measures of candidate **Utilities** / levers.

Since you don't yet know which factors will be most important in analysis, towards your above objectives, it is important to outline some candidates in a table like that below. If doing Predictive analysis, whether predicting a measured continuous or interval data, or training a classification engine on existing nominal groups, Utilities are going to include the predictors used in modelling (specifically they imply yet-to-be-determined weights on those predictors, which will come out of analysis). If you are doing Prescriptive (e.g., true Optimization with managerial outcomes), Utilities are the things you hope to change towards a managerial objective. If you are combining both methods, your listed Utilities may serve both purposes.}

Utilities	T/P/F	State (& Bounds)	Utilities	T/P/F	State (& Bounds)
	/ /			/ /	
	/ /			/ /	
	/ /			/ /	
	/ /			/ /	

Relational Connections:
{Insight into the impact that candidate Utilities / Levers, have on the Priority 1 (focal) Objective; I.e. **Connections**. This should include descriptions of any non-linearities, uncertainty, effect lags and possible O-I feedback.
A useful approach is use of a Relative-Impact Fishbone diagram. You might include views that suggest correlations (e.g., scatter plots) as well.

Root-Cause mapping: Relative-Impact (RI) Fishbones allow you to capture your early speculations on the potentially complex role of predictors in a visual form.

You may also find it useful to include a table, such as that presented below, to take note of complicated relationships between utilities, and how much they might be modified in the future (current state slack)}

Connections (and Utilities / Objectives involved)	Current State (& Slack)

Excellence Principle: {what mantra summarizes the prescription, goal and control plan?}

Labels along left margin (top to bottom): Define; Measure/Descriptive Analysis; Predictive Analysis

Figure 3.5 Left-Side Stage-Setting Foundation of the S-A3.

reference system, and descriptions of relevant Objectives, Utilities, and Connections that might be difficult to sufficiently describe within the confines of tables. For example, in providing details that help provide context for your lists of Objectives, notes provided in Background can benefit from the Why/Who/When/How questions outlined earlier, and in reference to outlining the current state and Objective candidate brainstorming described in Chapter 2.

However, importantly, this space is not unlimited, and for good reason. It's easy to say a lot about the dynamics, opportunities and challenges in a context. It's much harder to boil these down to what is truly critical. The space limitation here (again 'structure') forces a focus on only the most critical of issues. Dedication to the goal of presenting alternatives among candidate Objectives and Utilities is thus tempered to ensure that there is room left available for Plan B and Plan C, but while striving to ensure that additional candidates are not just adding noise in our considerations.

Our view within this frame can be further enriched, ahead of formal modeling efforts, through visual artifacts as well. Visuals can often be more efficient in conveying certain ideas than can words (more on this in Chapter 5). Since speculations at this point are chiefly focused on the nature of cause and effect, in an effort to think through and identify key attributes of our mental models of the project reference system, it makes sense to leverage other frameworks that have historically proven useful. One of these classic forms of cause-and-effect diagramming is the Fishbone (or Ishikawa) Diagram, attributed to Kaoru Ishikawa in his ground-breaking contributions to quality management in the Kawasaki shipyards of the 1960s. Little has changed in the way these diagrams have been applied over the intervening decades years (MacDuffie 1997; Nallusamy 2016), a clear testament to their general effectiveness. There are nevertheless opportunities for augmentation today. Since the magnitude of impact can differ greatly across prospective causes, and since issues such as Risk, Constraints, and Lags are so often underrepresented in mental models, let alone actual analytical models, it is would seem appropriate to draw greater attention to these issues if possible.

As it turns out, drawing attention to such points is very much within reach. An augmented form of the classic Ishikawa structure, which we refer to as a the Relative-Impact (RI) Fishbone diagram, was designed with enhancements to help visualize uncertainty, non-linearity potentially related to the influence of constraints, degree of impact, time-lags among Utilities, and a focal Objective of interest (Figure 3.6).

Among other options for creating such a depiction, a freeware open-source tool, the Blackbelt Ribbon add-in for Excel, is available at www. blackbelt-apps.com to facilitate these depictions. Let's talk briefly about the components of this type of depiction.

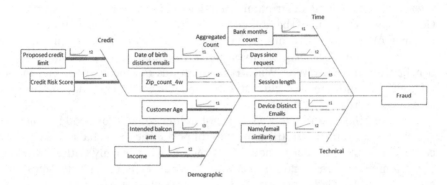

Figure 3.6 Example of a Relative Impact (RI) Fishbone Diagram.

a An Objective at the Head: This is the far-right node in the case of Figure 3.6, and will contain either a Fundamental Managerial Objective (FMO) or a Means Objective. All other elements in a single RI-Fishbone will flow into this node.

b Utilities as the Branch Nodes: Flowing into the Head of the diagram are all other nodes, categorizes by larger Branch groupings. These are generally our Utilities. In a predictive model, we are really talking about things like coefficients of effect associated with factors that predict the Objective. In a prescriptive effort, these are the real-world levers that might be modified in driving forward that Objective.

c Connections as the Branch Arcs: While the first two features are common to classical Ishikawa Fishbone diagrams, we are also interested in depicting details of the Connections between these. We want to depict these even before we have conducted associated analysis, to provide at least some strawmen to knock down (or perhaps fortify). We will also plan to go back and update these as clearer pictures emerge. The dimensions capture along these arcs include:

    i Form: Perhaps driven in part by things like constraints/bounds on either a Utility node or the Objective, what is the general shape of the given relationship? Do we expect a Utility in question has a strictly negative impact (the x-y is negatively sloped), is it positive, is it more complex over the range of values the Utility might reasonably take on. If the Utility in question is binary (True/False) or categorical, a visual might depict a step-function (ideally each category is broken out separately in advance). If the Objective itself is binary or categorical, as discussed and common when outcomes are constrained, the relationship between a continuous Utility and that Objective may be described by some asymptotic approaches, if not a step function as well (e.g., if there is a threshold in the Utility beyond which an impact is 'switched on'). In any event, the mini-graph on each branch arc allows for the communication of such details. In the freeware tool, our y-axis ranges from 0 to 1, to provide relative trend depictions across a five-equidistant value range of what can be viewed as typical, standardized, values in x.

ii Magnitude: Regardless of the shape of the x–y relationship (form), how much absolute bang does we think a minor (say 5%) change in one of the Utility nodes is going to have on the Objective? In combination with the shape of the relationship, this allows use to start to think about the total impact that the full set of Utilities might have collectively on the Objective. In the freeware tool, we scale magnitude, relative to other Utility impacts in the diagram, from 0 to 1.

iii Certainty (Risk): How certain are we that the impact anticipated will regularly play out? Even if we think a major impact might exist for a given Utility, maybe we are not all that confident that it will 'always' work out that way. This confidence of impact is also something we'll learn about as analysis progresses. In the freeware tool, we also scale certainty, in terms relative to the other Utilities, from 0 to 1.

iv Lags: Is the impact of a change in the Utility on the Objective immediate, or delayed? How delayed? It's all relative. If all impacts of all Utilities are really simultaneous, then things are somewhat straightforward. If difference lags exist, however, it's probably worth documenting these. You'll want to also consider such lags as you move into subsequent stages of The Cycle.

In the presence of clear feedback mechanisms, the diagram can be further augmented by the including of additional branches emanating forward from the Objective head. However, a perhaps more effective route will be the use of additional diagrams wherein the Objective of one serves as a Utility in another. This is also in the spirit of intermediate Means Objectives. The combination of multiple RI-Fishbones, with either serial and/or circular relationships can provide an effective roadmap as one checks of aspects of more holistic models they intended to build to capture the reference system in question.

Practitioner Recap

In The Cycle, Utilities are the levers that decision-makers have direct or indirect control over, and which have the potential to advance project Objectives. They are more than mathematical variables. Utilities must have the potential to impact an Objective, and their identification involves considering real-world constraints and evidence. The best Utilities demonstrate transparency (e.g., how well is the behavior of the Utility understood), plasticity (e.g., how much can the Utility be moved) and fit (e.g., how much organizational inertia exists to move the Utility). Connections refer to the cause-and-effect relationships between Objectives and Utilities. They help capture the real-world "process" in an Input-Process-Output sequence. It is important to view these Connections through a wide variety of perspectives, and to adjust mental models as evidence from analysis emerges. Risks, constraints, lags, and feedback mechanisms are key considerations in analyzing Connections. Challenging your intuition regarding connections via counter-argument efforts, and visualizing connections via concept mapping can be helpful brainstorming tactics for identifying and documenting factors affecting Connections. S-A3 documentation enables broader information sharing but, more importantly, it also creates a living history of key ideas, revealed evidence, and evolving alternatives that emerge during the project.

Chapter 4

Systematic Mental and Analytical Models

Embarking on this next stage of discussion, it can be useful to remind ourselves of the fundamentally systematic nature of the contexts in which projects are discovered and carried out, the multiple connections between solutions and various facets of those contexts, the importance of retaining a critical eye in assessing such solutions, and the value of being brave enough to admit that certain assumptions may need to be revisited. The following quotes help to keep us on task in this regard.

> *When we try to pick out anything by itself, we find it hitched to everything else in the Universe*
> *— John Muir (1911), My First Summer in the Sierra, p. 110 of*
> *Sierra Club Books edition*

> *Experimentation lies at the heart of every company's ability to innovate.*
> *— Stefan Thomke, Enlightened Experimentation, 2001*

From the start of our discussions, we have repeatedly referenced the concept of "systems" and thinking "systematically". We've also referenced some of the benefits of design thinking tactics which can augment brainstorming around structure, and alternatives central to The Cycle. While there are certainly notable champions of systems thinking, as well as stalwarts of design thinking, these two approaches really are far more complimentary than they are at odds. Consider one of the classic definitions of systems, with reference to systems thinking:

> *... any group of interacting, interrelated, or interdependent parts that form a complex and unified whole that has a specific purpose*
> *— D.H. Kim (1999), Introduction to Systems Thinking, p. 2*

DOI: 10.4324/9781003427650-6

Systems thinking, with its own host of tools and approaches to capturing details, focuses on understanding the interplay and dynamics among a system's parts, which collectively generate a unified whole. As in the bicycle example, the system depends on specification of the scope of the referents defining the system. Absent a rationalized scope, there is little chance of getting anywhere. However, this emphasis also shifts focus away from the mechanics of any single system component. Just enough needs to be known about each component to understand its role. In the vernacular The Cycle, the corollary to a reference system's scope are focal Objectives.

Further, The Cycle highlights aspects of the system (i.e., Utilities) that can in practice be influenced by decision-makers, thus advancing FMOs. The Cycle also emphasizes 'how' FMOs can be advanced, reflective of the dynamics characterizing the Connections linking Utilities and Objectives. A complete representation of core Objectives, Utilities and Connections within The Cycle, and a formal examination of that system as a whole is necessary to identify opportunities and challenges in pursuit of FMOs. This holistic approach facilitates further identification of potential, albeit unanticipated, biproducts of prospective solutions, and can prompt reconsideration of foundational assumptions.

How is design thinking different? In contrast to systems thinking, at least on the surface, it might appear that there is far greater emphasis on the identification of a solution for a particular user; and, only secondarily is there an emphasis on the mechanics underlying how a system encompassing, say, a problem-solution-user ultimately works. However, it is possible to argue that the mere specification of referents in a systems thinking exercise (e.g., the dynamics of a bicycle interacting with a rider, and a set of factors with potential to improve that experience) amount to the same thing. Nevertheless, in prioritizing a solution-user orientation, rather than a broader system dynamics orientation, design thinking has given rise to many successful innovation processes; again, assisted by a wide range of documentation tools and frameworks, although, candidly, the best solutions ultimately require a scientific understanding of local referents, including both mechanical and behavioral elements. Effective design thinking eventually leads to at least some degree of systems thinking in efforts to determine the underlying nature of problems and potential solutions. Likewise, the most practical approaches to systems thinking ultimately push toward the design of solutions that capitalize on a thorough rationalization of the nature of potential problems within a reference system.

The Cycle begins with a consideration of Objectives to prompt focus on a reference system for which solutions might be developed. As in design thinking, explicit emphasis in The Cycle is placed on arriving at such solutions. However, as in systems thinking, the Cycle highlights the complexity of Connections between Utilities and Objectives which can limit the utility – or even relegate to irrelevance – some initial inspiration. While the Cycle highlights the importance of alternatives, the momentum that structure can provide is also core to the value it provides. And, it is the holistic structure of what we know, or at least what we anticipate, that takes us into the final stages of the Cycle.

4.1 Manifest

As with the previous stages in the Cycle, a proper discussion of the Manifest stage begins with a relatively concise, but sufficiently inclusive, definition.

> **Manifest**, in the OUtCoMES Cycle framework, involves the assembly of the full set of Connections between Utilities and Objectives, towards the formation of an integrated system of relationships. The product is a system of interconnected rules and relationships that essentially define the scope of the decision space for analysis and design. It creates the foundation on which depend the exploration and search for solutions that fit both the problem specifications and the context for application.

Whether the focus is on a system of equations, or a rule and objective set to feed into a generative design program for product development; whether the goal is to develop a strong predictive forecast, or identify an optimal allocation of resources; or whether the work involves coding, or simply doing work with pencil and paper ... it is at this stage of the Cycle that formal assembly of all of the knowledge acquired up to this point commences. The aim is to build a cohesive and coherent representation that informs the local reference system.

Consider again the example of the bike and rider system from Chapter 1. At the Manifest stage, understanding of the distribution of likely human operators, which itself entails a range of inter-related attributes (e.g., weight, strength, endurance, etc.) is combined with knowledge of the inter-related attributes of the mechanical system (e.g., force required to rotate, subject to gear selection, desired speed, grade/incline, etc.). Common terms such as "force applied" and "weight" permit crossover, and ties between these human and mechanical characterizations.

As an alternate example, drawn from projects we've helped budding analysts develop in recent years, consider the complexity of studying hospital-acquired pressure injuries (HAPI). In such examinations, one

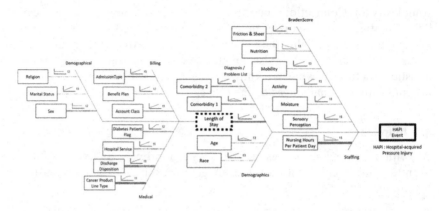

Figure 4.1 Serial RI-Fishbones in the Manifesting of a Reference System Model.

might begin with the presumption that attributes of the patient experience including the overall length-of-stay, might be predictive of the incidence of HAPIs. Such injuries, their prediction and ultimate reduction (or elimination!), may represent an FMO and conditional Analytical Objectives (e.g., maximize likelihood, accuracy, precision). The FMO also introduces focus on the prediction and reduction of intermediate outcomes, such as length-of-stay (a Means Objective). Ultimately, it may be possible to exert more control over the precursors of a Means Objective than over other predictors of the FMO. The prediction of these two Objectives might eventually occur simultaneously, in the estimation of a system of equations (e.g., part of the Explicate stage). In practice, however, it might also be possible to anticipate that efforts at estimating factors such as length-of-stay might have in fact already taken place; and that such a model's estimation is available from an alternative source (e.g., from operational efforts to manage capacity and bed availability). As depicted in stylized form in Figure 4.1, the two sets of relationships are clearly conceptually and functionally related.

If there is a strong suspicion that causal loops are present, these reciprocal interdependencies should be incorporated into the modeling effort at this stage as well. For example, it is reasonable to assume that actions leading to diminished length-of-stay subsequently also lead to a reduced number of HAPIs. These actions would undoubtably enhance the reputation of the health system (both among potential clients and investors), and potentially to increased demand for hospital services. Such an uptick could change the demographics of the clientele being served, and might introduce challenges to hospital service, with commensurate upward pressures on length-of-stay. It would be irrational to anticipate isolated changes in system Utilities to yield indefinite benefits to an Objective, even if these are evident in the short term. It is optimal to anticipate and capture such feedback loops conceptually at this stage. If there is reason to expect them to be present, their potential role should certainly be explicitly accounted for during analytical model estimation and the search for policy solutions.

At the Manifest stage, the constellation of relationships for which there is either prior analytical support, or which are substantiated by either contextual experience and/or theoretical rational are connected. Pragmatically, a fair bit of intelligent guessing is still necessary at this point – informed but not entirely confirmed. After all, if it could be known in advance how everything played out, analysis wouldn't be necessary. It is impossible to know in advance exactly how/whether the puzzle being built will yield a solution. But, assembling the pieces and creating a foundation for the estimation and/or search stage of the process helps to confirm or refute some of the ideas that have been developed thus far.

With such next steps in focus, development of sufficiently holistic reference system models, and the documentation of this effort also serve another purpose. The nodes and arcs in a series of RI-Fishbone diagrams, for example, can serve as a checklist as math concepts are converted into math or coded rules. Has the impact of Utility #4 on Means Objective #2 been estimated? Check. Has the predicted impact of Means Objective #2 on the focal FMO been captured in model calculations? Check. Has the fact that Utility #6 and Utility #7 are mutually exclusive Yes/No decisions, or are investments bound to a fixed pool of discretionary spending been accounted for? Check. Not only does this help guide the process of piecing together the puzzle, it also creates a foundation for sharing details of that integration with key stakeholders.

What kind of evidence would point to the coverage of all of the conceptual bases of such a model? Before attempting to extract intelligence from the model (Explicate), particularly in light of its likely inherent complexities, it is important to kick the tires a bit. Occasionally, this can be a straightforward exercise involving sensitivity analysis. Provide a series of scenarios for which the set of outcomes is known, ideally informed by real-world evidence of such outcomes, and see if the model cranks out the anticipated results. However, given that models are often built on estimated relationships (forecasts), there's bound to be some uncertainty regarding how scenarios can play out. Here, it is important to capitalize on empirical measures of that Risk. Simulation analysis may prove critical in assessing whether the Manifested model effectively captures the real-world reference system that the Objective(s), Utilities, and Connections are attempting to provide intelligence towards.

📟 *An Examination of a Manifested Model*

It can be hard to imagine what a simulated approach to ensuring the completeness of a Manifested model might look like, especially for those of you who may not have previously used simulation analysis. Fortunately, regardless of whether you are designing a new physical product and relying on tools like SolidWorks or Ansys, or trying to improve upon the features of an organizational environment, a wide range of sophisticated simulation options are available. As a case example, consider a common scenario with a fairly clear Objective and set of Utilities, but a somewhat complicated array of Connections. Complicated enough to potentially give rise to error in model assembly at the Manifest stage. In this instance, greatly stylized from the real-world case on which it is built, imagine an organization interested in making cost-effective decisions regarding the ordering of various products (for direct sale, or use in subsequent manufacturing or service).

For simplicity, we'll assume they are trying to come up with an ordering policy for five products. That policy will ultimately involve determining an inventory level for each product that triggers a replenishment order from the supplier, as well as how much extra to order in an attempt to hedge against various unknowns. We'll call that the re-order point and will assume distinct levels are possible for each product. Decisions will be driven in part by factors such as the estimated rate of demand (e.g., how fast the organization sells or uses these items), but also financial issues such as the cost of replenishment, the cost of running short (e.g., lost sales, expediting, or backordering), and holding (e.g., taking up valuable space). Regarding the latter, the organization also is trying to determine how much space to dedicate to storage. Storage space also implies some fixed cost, but not having enough space to accommodate stock on hand might incur additional rents. Since those can come at a premium charge, the organization wants to get the 'space' decision right as well. This is a lot to account for in making a series of decisions, particularly given that future demand (and lead time for orders) can at best only be guessed at.

In Figure 4.2, we see a rendition of the Manifested model. To maximize accessibility, we've built a version of this in MS Excel (downloadable from www.masteringdiscovery.com).

The workbook in question is operating in something referred to as 'iterative calculation' mode, permitting the functionality of circular references. In this instance, such a model of calculation allows the simulation to update system tallies (e.g., how many days have

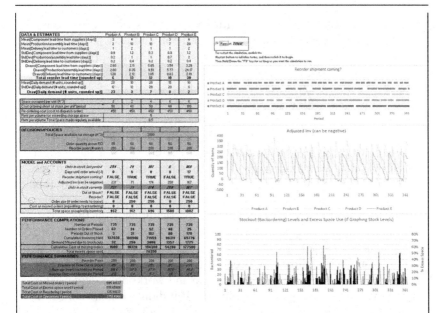

Figure 4.2 Snapshot of Manifested Model and Visualization of Simulated Dynamics.

transpired, on how many of those days have we run out of stock on a product, what is the current stock level, how much longer until the next resupply truck arrives, etc.). More detail on the construction of related models can be found in sources like Excel Basics to Blackbelt (e.g., Bendoly 2020). Critical for us at this stage is an assessment of whether sets of Utilities, Objectives, and Connections capture anticipated dynamics in support of desired solution development.

In Figure 4.2., and in a closer inspection of the workbook, the structure of the Manifested model is shown to consist of both reorder point and order buffer decisions, as well as the total space allocated for dedicated storage. Color coding here is useful, as with the broader application of the Cycle: Blue for Utilities, Green for Fundamental and Means Objectives, Lighter shading for more intermediate Means. In this case, the big prize is minimizing the total implied cost of operations (dark green bottom cell, including cost of space, shortage, ordering, etc.).

This will help serve as a check on model coverage. Any number of signals can serve as indications of the extent that the model is or is not capturing its intended features. For example, the dynamics of resupply shipments should be occasional, repeated, and broken up by at least limited periods during which orders are not being placed or enroute. The top right graphic depicts this dynamic for each of the five products (e.g., periods of waiting on vehicles represented by bars over a span of one year, with some white space separations). Another anticipated dynamic is a general ebb and flow of inventory in a form common to inventory systems of this sort; what is typically referred to as sawtooth dynamics. As trucks arrive with stock, the inventory levels for a product shoot up, followed by gradually diminishing inventory as it gets sold or used up. This seems to be the case for each of the five products as well, as depicted in the middle graphic on the right.

Prior to going through the process of searching for a more ideal ordering and space usage policy, it also is important to keep an eye out for issues with potential to become especially costly. In this system, these issues include excessive storage needs on the one hand and shortages (which may translate into backorders) on the other. Both cost money. In a multi-product setting, they might co-exist, but the chances of needing additional storage capacity will tend to be lower when stock shortages occur. The lower graphic on the right side of Figure 4.2 (and more clearly in the workbook) depicts this interplay, which emerges as anticipated. If it didn't, this would be an indication that a key dynamic had been overlooked.

These inquiries might not be entirely sufficient to ensure a fully functioning model of the system in question ... but it's a good start. Generally, the last thing you want to do is rush through the creation of a model full of errors, only to be left with flawed intelligence.

The assembly of a broad set of Connections, Utilities, and Objectives in a sufficiently comprehensive reference model can be daunting. There is certainly virtue in learning from examples, but still more in the possession of a roadmap ensuring that all relevant issues are covered. Tabular documentation of Objectives, Utilities, and Connections has proven to go a long way in this regard, as has the thoughtful construction of RI-Fishbones. The more time spent with this process, and the more examples where these tactics are used, the more facile engineers and analysis tend to become in tackling new areas of discovery.

4.2 Explicate

With a first cut at a holistic working model in hand, with a substantial area of the puzzle assembled, the stage is set to begin extracting intelligence solutions and finding a way out of the maze that the puzzle renders. This is the **Explicate** stage.

Explicate, in the OUtCoMES Cycle framework, involves the derivation of actionable intelligence, based on a set of understood
and interconnected rules and relationships. In predictive exercises,
this involves the estimation of effects and predictive structures
(i.e., predictive Utilities), whereas prescriptive exercises involve the
assignment of value to decisions (i.e., prescriptive Utilities). Heuristic
tactics, mathematical approaches, and computational algorithms
are all fair game, depending on alignment with the specific modeling
exercise in question.

While the task at this point can be fairly straight, forward, it can also be
extremely challenging, even with the most effective model in place. If the
task is relatively straightforward – i.e., estimating the effect sizes of
predictors in a forecast, or even computationally determining the structure
of a random forest or neural network – it can generally be handed-off to an
appropriate algorithm. While a model structure might yield an
unambiguous set of regressed coefficients in an OLS setting, the consideration of training and testing samples, the role of random elements and re-
examination are crucial in other machine learning contexts. Consideration
of the robustness of intelligence Explicated from this process is key to
moving forward.

The same applies to prescriptive tasks. In some settings, such as those
involving generative design and iteration between human-specified rules,
physical testing and AI-driven exploration and simulation, cycles of
search, discovery, and rule-adjustment are central. The same holds for
tasks that, though relatively simple from a modeling standpoint, create
an intense computational burden when it comes to extracting sound
recommendations. A classic example of a relatively concise model
structure giving rise to a grueling Explicate stage is the Traveling
Salesman Problem (or frankly most tasks that involve sequencing a
large number of items or activities). The challenge emerges from the
sheer number of possible sequencing combinations. If there are 20
customers in the queue, there are potentially 20 factorial combinations
to consider (or 2.4 followed by 18 zeros). For comparison, package
delivery services visit around 150 locations on any given daily route. Try
putting 150 factorial into your phone's calculator ... you might get an
error. On an Excel spreadsheet, you'll get about 5.7, followed by 262
zeros ... A daunting task, and generally not one that a comprehensive
search is going to help with. But, per Voltaire, we can't "let perfect be an
enemy of the good" (La Bégueule 1772). Sometimes, a 'good' way out of

a maze that emerges from savvy tactics ultimately offers a better fit to a problem or needs of a user than the 'best' way out, that can only emerge through brute force.

🚴 *Exercise*: Consider a simpler version of this problem with the 20 points mapped in Figure 4.3. For simplicity of illustration, a fair degree of geometry has been imposed here - specifically, what engineers call five-fold symmetry. The 20 points are arranged with each of five points 72-degrees apart on four concentric rings (rings only partially shown). For example, the points A, B, C, D, and E are on an outer ring. The distances between each neighbor on that ring is the same. There's also alignment between the rings. A line drawn through points C and M will also pass through the middle of the field (center of all rings). Similarly, a line drawn through points L and S will do the same. The two inner rings have radii of 5 and 10, respectively, and while the outer two rings have radii of 15 and 20.

That's a fair amount of detail that you may not have wanted, but it's relevant.

Imaginary rings aside, what if you were asked to draw the shortest path, starting at any point you choose, but which visited each of the remaining points, and returned to the point you began with?

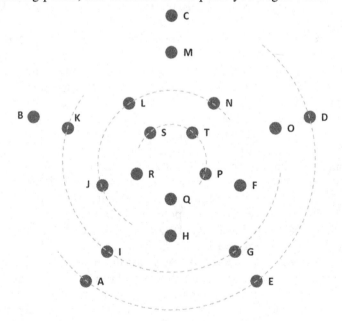

Figure 4.3 Twenty Points in Space.

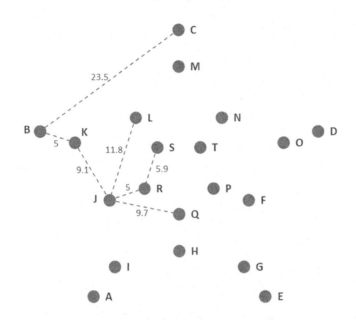

Figure 4.4 Twenty Points with Common Distances Shown.

The Objective is minimizing the total length of the line drawn. No other rules are imposed, apart from making sure that each point is on the path and that the path ends where it began. If you think that using curved lines are more ideal, you're allowed to (though, just as a heads up ... they aren't!). If you'd like to cross a path that you drew earlier in your journey, that's fine (though, just another head's up ... that's not going to be the shorted route). Just don't cheat by folding the paper ... we won't be creating Einstein-Rosen bridges in this setting.

To help you with this exercise, and to avoid the need to dig up a ruler, Figure 4.4 shows the same set of 20 points, with distances between some likely pairings. Remember that because of the high degree of symmetry in this landscape, distances between any two pairs of points in the same ring or between, are going to apply in repetition (e.g., the distance between K and J, of 9.1, is the same distance between K and L, L and M, M and N, etc.).

Still stumped?

Even with the information provided, and in part because of it, this is a problem that a lot of people struggle with. Just like a standard 20-location Traveling Salesman problem, there are still 20-factorial (20! = 2.43E18) combinations. Clearly, although many solutions are going to be lousy, the task is to find those that are outstanding.

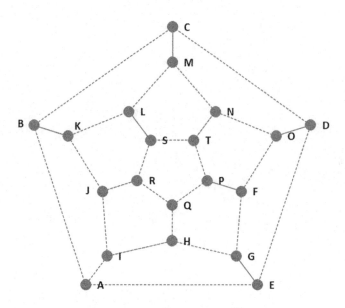

Figure 4.5 Twenty Points with Three Connections Each.

A little more structure might help to move the process forward. What if we just sketch out some elements that might be part of a good path, rather than trying to map out the entire path at once. In this case, let's sketch straight lines between each neighboring point on the outer ring, to make a simple pentagon. Now let's connect each of those five outer perimeter points to their next nearest point (B's would be K, C's would be M, etc.). Let's finish our sketch so that all points in the field are connected to EXACTLY three other points, selecting those pairing by order or shortest distance. The result should look like the dodecahedral graph depicted in Figure 4.5.

Now, what if you were told you that the shortest round-trip visit, through all 20 points, required nothing more than these drawn paths – and not even all of them? In fact, as you'll see, the best solutions use only 20 paths … ! What's happened is that imposing this structure significantly reduces the total number of options that need to be considered.

Think about this – you can still choose any starting point. But, your next choice is no longer from among 19 points. Instead, you only need to choose from among 3 alternatives. Moving forward, your next 13 or so choices will have at most 2 options apiece (i.e., having already used an incoming path). But, by the time you get to around the 15th point, if not earlier, your options will have been

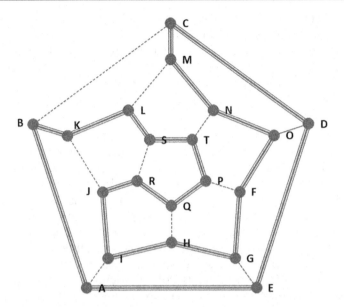

Figure 4.6 One of Many 'Best' Round-Trips: Total Distance 210.4.

funneled down to only one option for each of the remaining points until you get back to the beginning. That's something more like $20*2^{13}$ or 163,840 considerations. And, while this may still seem like an imposing set of options, in relative terms it is infinitesimal as compared with a set of 20-factorial options. It's also an overestimate of the number of complete paths that will allow you to return to the beginning (i.e., some of these are traps that lead to a dead end). One complete solution is shown in Figure 4.6.

Incidentally, this activity is an adaptation of the classic Icosian Game, developed by Irish mathematician William Rowan Hamilton back in 1857. The solutions to this game are referred to as Hamiltonian cycles. You could play this out on the edges and vertices of a dodecahedron (12-sided gaming die) as well. The flattened dodecahedral graph is just a lot easier to print on a page ...

What is perhaps most striking is how capable individuals are in solving this problem once the limited number of deliberately selected close-neighbor connections are drawn as in Figure 4.5. While some see the imposition of this structure as likely to limit the best options, in reality this approach simply helps to filter out poor ones. The structure makes the implications of stepwise choices transparent. And, because decision-makers often see problematic options looming, they tend to course correct, either returning to a prior step or starting over on a clean sheet rather than following their initial path over the proverbial cliff.

Structure yields insight. This axiom applies to all of the modeling we do, whether we get it right the first time (usually pretty unlikely) or whether the clouds part on the 12th revision.

Here, we can further capitalize on two key terms that continue to come up: structure and alternatives. They are clearly instrumental to success in the Hamiltonian Icosian exercise. Structure facilitates the consideration of alternatives. And, with structure, it becomes evident that a lot of different approaches could be used to solve the problem. After all, since a solution requires that all 20 points be crossed, you could start at any one of them and generate a solution. Looking at just one of the cost-minimizing solutions (e.g., Figure 4.6), it's also clear that you could move in any of three directions, from any one of the 20 points, and still generate a solution. Possible alternatives include 60 options for starting points alone, let alone all of the permutations for subsequent choices. Structure and experience set the stage for the emergence of great alternatives. And, as it turns out, a very similar expedient lies at the heart of all contemporary artificial intelligence algorithms.

Consider for example two common predictive tools in the AI/machine learning space: artificial neural networks and random forests. Both leverage the structural lattice of predictive computational node elements, typically arranged in sets of interconnected layers. It is the architecture of these nodes and their interconnecting arcs that are the Utilities in an estimation process (as in The Cycle) that can maximize Objectives like the accuracy and the precision of predictive classification and trend estimation efforts. Obviously, a huge number of alternative structures and node-arc configurations can yield a prediction. But in all cases some core structural choice underlies the effectiveness of the search for alternative models with potential to advance an Objective, and without which the search for alternative model estimations would simply be inconceivable.

In the prescriptive space genetic algorithms (GAs) represent a similarly illustrative exemplar. Genetic algorithms, as the AI/ML community seems to be rediscovering, can be an effective way to extract outstanding solutions from particularly challenging decision spaces. These include the generalized Traveling Salesman Problem (or any large-scale sequencing task). What makes GAs standout from greedy optimization approaches (e.g., SIMPLEX or GRG Nonlinear of Solver)? GAs begin with a sample of potential solutions (here, think 'possible routes' in the Traveling Salesman case), and then proceed to evaluate and cull poorly performing solutions. Remaining solutions can be paired for genetic crossover. In a sequencing case, this could involve a mini-sequence of five stops within a larger route being swapped for an alternate (relative) sequence of those same five stops in an alternate route solution. Although it's a bit more nuanced, GAs also capitalize on random mutation of solutions to draw in 'new genetic material.' But central to any GA approach is the clear articulation of both structure and alternatives. The evolutionary process that GAs capitalize on encompasses tangibles like mini-structures within solutions. To build new and better alternatives, these then serve as a pool of next-generation structures which allows propagation of the search process.

Highly accessible technical demonstrations of the use of GAs can be found in sources such as Excel Basics to Blackbelt (3rd edition). While tools like Solver have small evolutionary engines built in, as alternative search mechanisms, more industrial-level Gas also have long been available (e.g., the Palisade suite's RiskOptimizer resource). However, regardless of the analytical tool applied in the Explicate stage, it is critical to not underestimate the importance of commonsense applications. Well-articulated model structure emphasizes the value of work done in all prior stages of the OUtCoMES Cycle. After all, fast and frugal heuristics have long served practice, and certainly prior to ubiquitous modern computing (DeTreville and Browning 2023; Gigerenzer et al. 1999). Many are still highly effective today, including one referred to as Nearest-Next. Applied to the Traveling Salesman problem space, it involves picking a starting location and simply selecting each subsequent location based on the next nearest alternative. This certainly doesn't guarantee optimality, but it can be highly effective, even if only as a starting point for GA search (e.g., Excel Basics to Blackbelt discussion and tool access at www.blackbelt-apps.com). When comprehensive searches of a highly complex decision space are untenable for the Explicate mission, combined approaches encompassing the use of heuristics and supplemental computational search can be highly effective.

🌎 Look Before You Leap

The business analytics team of a large multinational organization welcomes Kenzie as their newest ML (machine learning) engineer. After onboarding, the group immediately tasks her with building a propensity model. The intent of the model is to determine whether any given customer will click on a particular product featured within the body of its distributed online marketing content.

Kenzie's first step is to build a reference model. She decides to use a set of training data gathered by her previous team; data which was used to solve a similar problem in the recent past. This training set primarily included data from two categories of products, each of which had been heavily promoted. However, her new team is focused on marketing products in different categories, with attributes that are distinct from those found on products in the training set.

Failing to recognize this important difference, Kenzie moves forward to validate the model using product data specific to her current task. She discovers only at this point that the associated data misalignment (a form of leakage) must be compromising the efficacy of her model. She goes back a few steps, makes more appropriate elections in her training data, rebuilds, and then retrains her model. Due to time pressure, her only option is to purchase suitable training data from an outside vendor; another realization, though a step that would be difficult to avoid in any case given the time allotted to build the model. This redo adds additional marginal costs to the project, but it is far more cost-effective than attempting to disregard or force through a poorly fitting model.

Kenzie's revised model, built with new training data, ultimately proves successful. However, she is left distressed, and somewhat self-conscious as a result of her initial missteps.

Learning Concept: **Availability Bias** – Inordinately allowing recent events or experiences to influence one's approach to a decision; E.g., influencing one's approach to the Manifest and Explicate stages of analysis.

Reflection Questions

1 What prompted Kenzie to move too fast and prematurely into model estimation?
2 What role did availability bias play in this situation, and how could Kenzie have avoided or at least minimized the effects of this bias?
3 What advice would you give to Kenzie moving forward? How should she view the cost of returning to previous stages of consideration, relative to that of inferior results? Can considering these costs help her value work done prior to model estimation?

Research Follow-up – Find out more about availability bias and research the steps in model assembly. What are some tell-tale signs of falling into this trap?

4.3 Scrutinize

What emerges from the process of extracting intelligence in the Explicate stage is often nothing more than one option among many. It may be that many other options are not immediately obvious, due to limitations, rationalized or otherwise, at previous stages. Following the Explicate stage it is essential to apply a highly critical eye as to whether derived solutions represent both a fit to the originally conceived problem as well as a fit to stakeholders who would benefit from the solution.

> **Scrutinize**, in the OUtCoMES Cycle, involves closely and pragmatically examining outputs of the prior stages, prompting corrective redress and reconsideration of alternatives when fit between the problem, solution and user is neither obvious nor sufficient. Although positioned at the end of The Cycle, more realistically, consistent with their respective descriptions and the tactics recommended for executing prior stages, scrutiny pervades each stage.

It is important here to emphasize that positioning Scrutinize at the end of The Cycle is deliberate. The aim is to situate Scrutinize as a gatekeeping mechanism. However, it is equally important to emphasize that scrutiny has been applied to all stages of The Cycle. In Chapter 2, candidate Objectives were scrutinized using a range of evaluative criteria. Such scrutiny facilitated improvement in the prioritization of FMOs, and rationalization of Means Objectives and Analytical counterparts. Similar scrutiny was applied in the examination of Utilities. The consideration of levers available in pursuit of Objectives might include high levels of Transparency, Plasticity, and Fit. In the discussion of Connections, baseline assumptions regarding the relationships between Objectives and Utilities were scrutinized, emphasizing whether insufficient consideration was being given to uncertainty and risk, constraints and non-linearities, lags and feedback mechanisms. Such scrutiny led to the encapsulation of these complications in visual form (i.e., the RI-Fishbone structure). Whether these elements and relationships were sufficiently accounted for and integrated into the full Manifest of a local reference system model, in consideration of examination tactics such as simulation, was then scrutinized. And now, with respect to a range of analytical tactics, a close look has to be taken at the intelligence extracted. Does it make sense? Does it offer a fit to the problem? A fit to stakeholders? To users embedded in the context where the solution would be applied?

In this latter role, it is essential to remain open to the insights Scrutiny can yield. Even the best, most representative results can be surprising. They can also be disappointing. And sometimes they can highlight issues overlooked in the larger process or assumptions that unnecessarily constrain the search for good solutions. Errors of omission can yield inflated performance expectations, whereas errors of inclusion can do the opposite. Both can lead to less-than-ideal solution recommendations. Sensitivity analysis applied to various aspects of a model during the Explicate stage can help highlight impediments like binding constraints on extracted solutions. Comparisons of extracted solutions against adjacent solutions in the 'same neighborhood' (i.e., differing only marginally along subordinate dimensions), and with an eye on the limits of those dimensions, can inform this process. If the results of sensitivity comparisons align with intuition, this can suffuse confidence in the fit of the solution. It might also be possible to identify ways to modify the model, focusing on fit with user needs, and bringing into focus issues constraining solutions and performance. This process can play out in the immediate term, or the longer term with emphasis on continuous improvement in the reference system's context.

Such adjustments explicitly introduce the issue of timing. On the one hand, of course, it would be optimal to put functioning solutions that add value in the short term into practice earlier than later. On the other hand, implementing solutions also can freeze the potential for subsequent improvements in the intermediate term. If you have project experience, you've likely found that 'freezing' can be driven by a number of factors. Project personnel may have been reassigned, thus being temporarily unavailable for further continuous improvement efforts. Top-level management, or clients, might not be interested in these efforts, being satisfied by the last set of solutions and results. Marketing may have run with specifics of the artifacts developed, and thus a service or product lifecycle clock might now be in place. Motivation to engage in continuous improvement, in the absence of a strong corporate culture, also may generate hurdles (we'll talk a bit more about this in later chapters). Suffice to say, identification of a finish-line can be challenging, at least in part due to the long shadow engendered by that choice. However, if final scrutiny of solutions for fit to problems/stakeholders/users, as assessed through sensitivity analysis, suggests that only relatively marginal added value can be squeezed out of the project in the short term, then it's probably time to move on. Before that point, however, keep the Cycle moving.

4.4 Documentation and Alignment

Having now discussed these final stages of The Cycle, it is useful to return to the discussion of best practices in documenting ideas and experiences that emerge from these stages. The Manifest, Explicate, and Scrutinize stages are clearly related to, but also functionally distinct from, the

Objective, Utilities, and Connections stages. They even sound different. The first three stages sound like nouns. Most of the work in those stages is descriptive, although the potential for prediction and prescription bear on assessments of which details to focus on. The latter stages sound more like verbs. In these stages, the focus is on taking action, typically with the assistance of statistical methods and computational tactics and resources, largely with predictive and prescriptive modeling, estimation, and search activities in focus.

Documentation of actions encompasses the description of how specific elements are systematically involved in such actions. As a consequence, documentation will tend to capitalize less on speculative and descriptive tables and concept maps, and more on specific steps taken and derived findings, in the form of verbal descriptions and graphs capturing predictions and prescriptions.

The right-hand side of the S-A3 is designed to house such details. It also prompts project personnel to further reflect on their expectations, i.e., 'future state', regarding the solutions developed (including risk around expected outcomes), and provide considerations for moving forward with continuous improvement opportunities (i.e., with perhaps already forgone alternatives), or in anticipation of future shocks that might inspire next-generation development projects in this context (see Figure 4.7).

Critical in this latter portion of documentation is a closing of the loop. Specifically, the documentation must shift from analytical evidence back to managerial relevance. This begins with a discussion of the Explicated details of derived from actions such as model fitting, or optimization, and rounds off with speculations regarding future steps and states of practice that might emerge. The following two issues must be highlighted at this stage:

1 What objective empirical evidence can you provide to suggest that advancements towards your FMO (from your left side of the S-A3) have been achieved? In other words, what are the specific improvements to your corresponding analytical objectives? If you were unable to provide improvements, that's certainly going to be disappointing, but can be informative for future efforts.
2 What were the key Utilities and Connections which seem to have been the most instrumental in this advancement? Which accounted for the most variance, or provided the greatest leverage in solution development? Were there any special features (e.g., lagged, constrained nonlinear, etc.) that describe these Connections for future reference? What do these imply regarding Prescriptive future action?

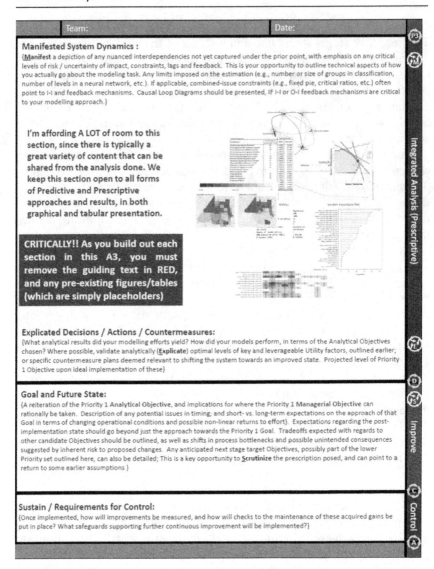

Manifested System Dynamics :
{**Manifest** a depiction of any nuanced interdependencies not yet captured under the prior point, with emphasis on any critical levels of risk / uncertainty of impact, constraints, lags and feedback. This is your opportunity to outline technical aspects of how you actually go about the modeling task. Any limits imposed on the estimation (e.g., number or size of groups in classification, number of levels in a neural network, etc.). If applicable, combined-issue constraints (e.g., fixed pie, critical ratios, etc.) often point to I-I and feedback mechanisms. Causal Loop Diagrams should be presented, IF I-I or O-I feedback mechanisms are critical to your modelling approach.}

I'm affording A LOT of room to this section, since there is typically a great variety of content that can be shared from the analysis done. We keep this section open to all forms of Predictive and Prescriptive approaches and results, in both graphical and tabular presentation.

CRITICALLY!! As you build out each section in this A3, you must remove the guiding text in RED, and any pre-existing figures/tables (which are simply placeholders)

Explicated Decisions / Actions / Countermeasures:
{What analytical results did your modelling efforts yield? How did your models perform, in terms of the Analytical Objectives chosen? Where possible, validate analytically (**Explicate**) optimal levels of key and leverageable Utility factors, outlined earlier; or specific countermeasure plans deemed relevant to shifting the system towards an improved state. Projected level of Priority 1 Objective upon ideal implementation of these}

Goal and Future State:
{A reiteration of the Priority 1 Analytical Objective, and implications for where the Priority 1 Managerial Objective can rationally be taken. Description of any potential issues in timing, and short- vs. long-term expectations on the approach of that Goal in terms of changing operational conditions and possible non-linear returns to effort}. Expectations regarding the post-implementation state should go beyond just the approach towards the Priority 1 Goal. Tradeoffs expected with regards to other candidate Objectives should be outlined, as well as shifts in process bottlenecks and possible unintended consequences suggested by inherent risk to proposed changes. Any anticipated next stage target Objectives, possibly part of the lower Priority set outlined here, can also be detailed; This is a key opportunity to **Scrutinize** the prescription posed, and can point to a return to some earlier assumptions }

Sustain / Requirements for Control:
{Once implemented, how will improvements be measured, and how will checks to the maintenance of these acquired gains be put in place? What safeguards supporting further continuous improvement will be implemented?}

Figure 4.7 Right-Side Model Assembly and Solution Extraction Stages of the S-A3.

Being able to outline these details with language specifically targeted to the relevance of findings, helps emphasize the value of the project effort in a way that might otherwise be hidden to a wide variety of stakeholders. It forces the project team to make sense of the work they have done, and in so doing facilitates follow-on projects in the same domain.

4.5 Looking Back on the Process

With a full view of the OUtCoMES Cycle, and the potential for leveraging structured documentation to highlight ideas and alternatives, it also is useful at this point to take stock of things. It can be valuable, for example, to take a moment to reflect on how The Cycle lines up with the core principles and aims of other associated value-adding flywheels, as originally referenced in Chapter 1. Table 4.1 provides a comparison of the stages and foci of The Cycle relative to a number of alternatives.

A few points are worth noting here. First, these flywheel cycles all have a great deal in common, despite the use of alternate naming conventions, and emerge from notably distinct contexts. The concept of continuous improvements embedded in PDCA and DMAIC is certainly central to The OUtCoMES Cycle, as evidenced by the emphasis placed on documenting alternatives and ensuring a structure amenable to cycling back when scrutiny warrants recalibration (c.f. Pyzdek 2003; Project Management Institute 2017). At the same time, similar to Design Thinking frameworks, The OUtCoMES Cycle clearly privileges opportunities for more immediate amendment between more local stages (i.e., not simply waiting to see a final recommended solution extracted).

It is also worth noting that, again similar to Design Thinking, reference to such terms as "Act" is deliberately excluded in The Cycle. There are at least two reasons for this. First, from a holistic perspective 'Act' is implicitly encompassed by Scrutinize stage. The intelligence derived from iterative evaluation is nothing doesn't advance either subsequent inquiry or application. In the case of subsequent inquiry, the ability of core Objectives to be rationalized and repositioned as Means to a next-level FO is Scrutinized. And so The Cycle continues. In the case of field application for example, when putting a product or process into action in the marketplace, the Scrutinize

Table 4.1 Comparing the OUtCoMES Cycle to Complementary Fly Wheels

	OUtCoMES	PDCA	DMAIC	Basic Double Diamond	Design Thinking "6" + 1 D-D View
Stages	Objectives	Plan	Define / Measure	Discover / Define	Understand* / Observe / Define
	Utilities				
	Connections				
	Manifest	Do / Check / Act	Analyze / Improve / Control	Develop	Ideate / Prototype
	Explicate			Deliver*	Test
	Scrutinize				Reflect
Objective Comparisons	Emphasized			Implied	Implied
Managerial vs. Analytical Forms	Emphasized	Implied	Implied		
Documenting Alternatives	Emphasized			Implied	Implied
Local Feedback / Revision	Emphasized		Implied	Emphasized	Emphasized
Continuous Improvement	Emphasized	Emphasized	Emphasized	Implied	Emphasized

Notes
* Stages such as Understand and Observe are often folded into the concept of Empathy.

stage extends across a window over which lagged performance must be measured.

This facet of the applicability of the Scrutinize stage in turn emphasizes the second reason why a distinct "Act" is not explicitly referenced in The Cycle. In the spirit of continuous improvement, and with a formal acknowledgment that solutions can inspire still 'better problems', it is important to deemphasize an implied terminus. Progress should be valued, intelligence should be developed, and even Objectives accomplished. But, it is also incredibly important to revel in the opportunities that these efforts provide. The pursuit of Objectives should always be viewed as way to open new doors rather than as a way to close old ones.

However, again, there is far more overlap than there is contrast among these approaches. The OUtCoMES Cycle is more explicit regarding specific steps, but its structure nevertheless critically values the core principles on which these other frameworks are predicated. The aim here is not to suggest that any one approach is superior for project work, but that the most impactful problems and associated solutions emerge from the joint consideration of multiple vantage points and the use of combined approaches. As highlighted in the next chapter, a wide range of tactics can be applied to data analysis throughout the stages of The Cycle (or any flywheels). Appreciating these options, and having a structure within which to consider alternative tactics, is as important as committing to best practices in problem discovery and solution development.

Practitioner Recap

Systems thinking requires a focus on the purpose underlying the whole project, while considering how the parts of the project contribute to its functioning. Because the interaction of these parts drives functionality, understanding the role and quantifying the impact of each part is vital for development of an accurate model of the system. However, trying to review the interconnected rules and relationships in their entirety is an exercise in futility for anything but the simplest of systems. Imposing structure to limit the factors under consideration at any one time will reduce the risk of chasing an impossible level of precision at this stage. Connections between smaller collections of nodes within the system provide a means of assembling this broader structural picture of system dynamics. Although such a system model is directly informed by the Objectives, Utilities, and Connections, it still requires empirical validation. It is critical to assess this previous work to increase confidence it informs a sufficiently credible model that adequately represents the real-world system. Close and pragmatic scrutiny of the outputs from the prior stages, with reconsideration and appropriate implementation of alternatives is required when fit between the problem, solution, and user is neither obvious nor sufficient.

Part II

Orchestration

Chapter 5

Synergies in Numerical and Visual Tactics

As we have emphasized, the effectiveness of The Cycle (or any project framework and process for that matter) relies on an openness to various perspectives. Accordingly, it also implies an openness to various approaches to analysis, and a potentially wide range of skillsets, all of which must work in concert if progress in key Objectives is to be advanced. The value of integrated 'combined arms' approaches to discovery and project development is common in quotes from recognized innovators, such as these:

Synergy means behavior of whole systems unpredicted by the behavior of their parts taken separately.
– Buckminster Fuller, Synergetics: Explorations in the Geometry of Thinking, 1975

If I have seen further it is by standing on the shoulders of giants.
– Sir Isaac Newton, 1675 (letter to Robert Hooke)

The OUtCoMES Cycle is a structured process, which encourages the development and documentation of alternative points of focus, alternative presumptions of cause and effect, and alternative solutions to identified problems. The process and structure is approached in a deliberate, informed, and step-wise fashion, with continuous opportunities for recursion as evidence presents itself. New ideas, findings, and interpretation inspire logical flow, regardless of whether we are moving forward in the process or returning to prior stages of consideration. To that end it relies on coordination across multiple methods and ostensibly multiple actors and stakeholders, an input/output, managerial/analytical communication dynamic which S-A3 documentation is specifically designed to promote.

Synergy among methods applied at each stage is crucial. By extension, coordination among those skilled in such methods, and charged with

DOI: 10.4324/9781003427650-8

executing and reporting back on the results of their application, is similarly critical. The best bag of methods and technicians amounts to nothing if the pieces don't fit together meaningfully. Simply recognizing the necessary methods and talent is also not enough. To identify the best problems, and derive the best solutions, projects need thoughtful, synergistic coordination across methods and people.

⟨⑤⟩ *It takes an Analytical Network*

Imagine a scenario in which an organization finds itself facing down the market-threat of a technology designed by an outside competitor. To fully appreciate the nature of the threat the organization, which deals with a myriad of concerns on a regular basis, needs to place the task in the hands of a competent and dedicated project manager. Appropriately, the adroit project manage begins by recognizing the fundamental need for understanding how the technology in question works. He begins by assembling a team, a network of field techs, with the express mission of data collection. Their task is to observe the technology in use, and construct a picture of the "who's", "when's", and "where's" of the technology's main application and purported threat. Through this Descriptive approach, the project manager comes to understand that the competitor is in fact leveraging the technology in partnership with still other collaborators, in what seems to be a kind of pincer maneuver against the organization's market position.

Hardened, and with support of the organization, the project manager poses a candidate FMO: Undermine the market position of the competitor by using their own technology against them. The analytical corollary to this very Prescriptive objective might be "maximize cost to competitor" or something of the sort, with measurement relative to some theoretical limit (e.g., insolvency). While sounding extreme, it aligns well with the relationship that exists between the organization and this particular competitor. In any case, the means of getting to any prescription that might support this call still must involve Means Objectives like "determine how the technology works." In this latter case, Analytical Objective corollaries include maximizing the accuracy and/or reliability of a model of the technology's functionality. A Predictive task. To that end, the project manager begins an effort to reverse engineer the technology in question. Once again, he draws on methodological expertise suited to the task, an expert in predictive modeling.

The predictive effort initially faces considerable uncertainty. It isn't immediately clear, even to the expert how to best approach the development of a model. The technology itself, it turns out, is in a constant state of redevelopment specifically intended to make this effort difficult. Still, the modeling expert has some ideas. In this instance, he draws on the best machine learning algorithms to tackle the task of disentangling causal, predictive Utilities and their Connection to the technology's output. Eventually, after identifying several crucial connections, a sufficient level of understanding is gained (nearly 100% accuracy). Through conversations with the project manager, and documentation of methods and findings, the modeler conveys how the competitors are using the technology. As a result, the organization, with guidance from the project manager, begins to use the technology defensively, nullifying the aggressive moves.

Prescriptive analysis, focused on the FMO, and leveraging the predictive model, can now finally begin. The project manager identifies an opportune point in time, given the organization's own resource availability relative to that of the main competitor, suggestive of an offensive strike of their own. The manager, in coordination with the modeling expert and network of field techs, deploys their own version of the technology, catching the principal competitor off guard. This move allows the organization to reinforce its own market position, and capture much of the position previously held by the competitor. Eventually, the competitor, having put all of its eggs in this proverbial technological basket, exits the market permanently.

This story isn't merely a description of a hypothetical journey of analytical coordination and synergy across organizational actors. Something very much like this actually happened 450 years ago in England. At the time a political conflict existed between the court of Queen Elizabeth I (aka "the organization") and Mary, Queen of Scots (aka "principal competitor"), who had not too secretly aligned herself with the French in an effort to take the English throne. Mary and her collaborators had developed a cipher for encoding messages to the French, with such an end in mind. Elizabeth's principal secretary, Sir Francis Walsingham (aka "project manager") was made aware of these likely correspondences. He employed a network of spies and double agents (aka "field techs") including Gilbert Gifford, to observe and intercept samples of the coded messages. With a sense of the lay of the land, Walsingham then leveraged an expert linguist, cryptologist, and forger, Thomas Phelippes (aka "modeling expert") to decipher and set the stage for defensive and offensive maneuvering. Eventually, the timing was right for a final strike. The so-called Babington Plot involved an encoded message from Mary with an implied order for her collaborators to assassinate

Queen Elizabeth I. The letter was intercepted, decoded, and used as primary evidence for the trial and execution of Mary, Queen of Scots.

Communication and coordination in the identification and solution of a problem are instrumental to the effective use of highly capable methods and actors. Those operating with such synergy tend to gain the upper hand. Those who are not quite as diligent can face far less appealing outcomes.

5.1 Coordinating Analytical Methods

In Chapters 6 and 7, we will dedicate our time to discussions of how best to play the project manager role in all of this. Beyond any basic training that you may have had regarding the use of project management software, the use of tracking and planning approaches and tools, effective Gantt charting, the best managers of analytical and engineering projects are highly capable of managing people. Specifically, fostering coordination and communication to enable the kind of synergies that efficiently evolve good projects into great ones, maximizing returns on time and resources invested. In short, we'll be focusing on strategy and tactics in managerial organization.

Ahead of that, it is useful to provide a corresponding bridge in terms of the array of analytical tactics that will characterize the varied expertise of individuals on a project team. These analytical tactics ultimately support the coherent and strategic use of the OUtCoMES Cycle, and therefore appreciating the related but distinct nature of these tactics is critical. These distinctions will also help us coordinate the efforts of individuals on our project teams, given that their expertise and availability will need to be leveraged deliberately to fill in gaps in our understanding. An insufficient understanding of the range of analytical tactics available at each stage of project work typically translates into an insufficient understanding of the skillsets possessed by a team. The result of insufficient tactical understanding tends to be bottlenecks and delays, misinterpretation and prescriptions that are poorly aligned, or simply erroneous. Adequate understanding of the range of analytical tactics available supports the best management of projects.

One useful way to categorize analytical tactics is according to their use in Descriptive, Predictive, and Prescriptive efforts. We've been discussing these genres of analysis throughout the text, with considerable attention given to their definition as early as in Chapter 1. We've learned that the OUtCoMES Cycle can be adopted with a focus on any one of these (e.g., Figure 1.5), but have also seen in later chapter discussions how the Cycle can be used to evolve understanding from Descriptive to Predictive and on to Prescriptive stages of consideration. Descriptive tactics set the stage for Predictive modeling. Predictive models provide a foundation (and sunlight) for developing Prescriptive solutions in the presence of otherwise uncertain structures and features of cause and effect.

But what are some of the specific examples of Descriptive, Predictive, and Prescriptive analytical tactics (i.e., computation and statistical methods) that can be leveraged for these tasks, and around which individual team member expertise might orbit and develop? In the 2019 whitepaper "A Framework for Analytical Approaches", written for the International Institute for Analytics (IIA), a wide variety of available analytic tactics for Descriptive, Predictive, and Prescriptive efforts is presented as a comprehensive set of interconnected decision trees. That is, analytical methods, computational or statistical, should be justified by the questions being asked and the nature of the analytical context. You don't use a hammer to bore a hole in a metal sheet, even if that's the tool you're most familiar with and happen to have available. You identify the correct tool. If you don't have it, get it. Or give the task to someone else who can do it and find something that you and your tool set can tackle. Misalignment seldom yields anything good, and not bothering to acquaint yourself with what tools others have used, and which tools best fit needs, is certainly a recipe for misalignment.

While we will not recreate the analytical decision trees presented in that whitepaper here, we'd strongly recommend that those interested reach out to IIA for copies. Instead, is it sufficient for our interest to discuss the nature of the methods that fall into each of the Descriptive, Predictive, and Prescriptive domains.

5.1.1 Tactics Used in Descriptive Analysis

Referring back to the definitions of these domains provided in Chapter 1, the kinds of methods associated with Descriptive efforts should be familiar to most. They are, after all, intended to describe the lay of the analytical and engineering landscape, stopping just shy of prediction; albeit typically proving highly suggestive to subsequent Predictive and Prescriptive efforts. Descriptive methods, beyond efforts to clean and organize data (investments in time that are certainly not to be under-estimated), include such basic practices as examining, describing, and approximating the distributions of observed data. At the simplest level, this can involve calculating measures of centrality such as averages and medians, and measures of variation such as standard deviation. However, the efforts should also involve examination of the shape of distributions, and efforts to test for whether the data fits specific know distributions. Is the data Normally distributed, or is it better character-ized as exponential, or something else. Would it be more meaningful if a transformation (e.g., taking the Log of the data, Yeo-Johnson transformation, etc.) was applied? Certain subsequent methods might depend on assumptions regarding these distributions.

In an associated sense computationally and statistically supported aggregation tactics can also prove useful for descriptive purposes. If you have seventy data fields coming out of sensors or surveys, are all of these truly providing unique details, or are many of these simply observable proxies for more latent factors that can't be directly measured? If that is the case, it might be more meaningful to aggregate multiple fields (think "columns") into a single factor for subsequent analysis. Numerous factor and component estimation techniques exist for such purposes, including exploratory factor analysis, principal components (PCA), axis factoring, and assorted applications of Bayesian techniques. The performance of these methods typically are often gauged by the degree of total and incremental factor variance accounts, as per calculations of eigenvalues (Analytical Objectives for these efforts).

Similarly, just because you have millions of observations doesn't necessarily mean that the individual nuances captured by each are of interest for the analysis of management policy and engineering applications. It may be that grouping observations, and considering the distinctions within and between groups, may be more critical. If you don't have a grouping schema already in mind (e.g., if there isn't already a field capturing "group" designation), exploratory clustering can be a very useful descriptive tactic. What you get out of these efforts is a sense of how observations (think "rows") fall together in ways that might not otherwise have been anticipated. Specific methods to this end include k-means approaches, hierarchical agglomeration, and mixture models. Analytical Objectives helpful in demonstrating the effectiveness of such Descriptive efforts include the pursuit of high ratios of between (inter-class) to within (intra-class) variation.

The individual measures of centrality, variance, and distribution can further be used to describe derived factors and clusters. Implicit in the creation of factors and clusters is the use of computational and statistical tactics for ensuring that things that are similar get joined together. However basic variable relationship examinations (e.g., correlations, bivariate distribution examinations), and group comparisons, such as T-Tests and ANOVA, can provide highly suggestive insights as analysis transitions into more Predictive efforts. To be absolutely clear, however, exploratory cluster analysis is explicitly a Descriptive method. It tells us what/where things might have been or what/where they might be now. Just like the calculation of an average. If the clusters derived from exploratory methods are found to be meaningful for future applications, then the question of predicting placement into clusters, or establishing a rule-structure for future classification can take on a Predictive or

Prescriptive aspect. Similarly, simple statistical comparisons of groups of observations may describe important differences and can be suggestive of Predictive and Prescriptive efforts. For example, you might find components described by a general Utility classification (e.g., metallic, high-quality, or Canadian) to be significantly different in a particular Objective performance measure. That might imply that future items of the same classifications perform with similar distinction, however, the leap being made is from something that is merely correlational to one that is far more causal. Such speculation can be insightful but falls short of formal Predictive (understanding cause) or Prescriptive (designing cause) analysis.

5.1.2 Tactics Used in Predictive Analysis

Predictive analytical methods capitalize (rely) on understanding obtained through Descriptive efforts. In the OUtCoMES Cycle, we emphasize the importance of identifying candidate Objectives that meet sufficient levels of Transparency, Plasticity, and Fit (see Chapter 2). If you are unable to provide analytical descriptions of the Objective you would like to advance, there's little hope that you will be able to predict or prescribe solutions related to it. If the description of the Objective demonstrates little variation, or identifies that the Objective has reached an upper limit, similarly, there is little point in attempting to advance it. However, if there is room to grow, and that growth is aligned (fits) the interest of the organizational mission, key stakeholders, and users, then efforts to identify Utilities toward that end can be useful. In the same sense, however, our initial consideration of candidate Utilities is a descriptive one. If a Utility hasn't changed much in the past, it might not other much in the way of Objective prediction. In other words, Predictive efforts benefit from all of the deliberate Descriptive work that we do with candidate Objectives and Utilities.

At the same time, Predictive efforts are often the key to realizing increasingly meaningful FMOs, by way of providing additional measurable Analytical Objectives (e.g., R-square, log-likelihood, accuracy, and other goodness of fit measures). The specific objective measurement of Predictive performance of course depends on a number of things, the chief being the nature of the phenomena / metric being predicted. In the IIA decision tree, for example, distinctions are made between efforts to classify (either into a range of nominal categories or into one of two binary classes), and efforts to predict ordinal or continuous interval phenomena. That is, OLS linear regression, while convenient (e.g., LINEST in Excel, Linreg via the Blackbelt Ribbon add-in, or any number of alternatives), clearly isn't the right methodological choice for a wide range of scenarios.

If the task involves classification, support vector machines or any number of forest estimation tactics can prove highly effective. If the data is nominal but ordered, such as levels on a Likert-type scale, ordinal classification tactics can be applied for further advantage in fitting. Alternative tactics can be applied when the dependent variable is [0,1], or if two classifications exist. In such cases, logistic regression (also made convenient by the Blackbelt Ribbon), neural networks or Bayes point machines can prove suitable. As with most machine learning (ML) applications, model estimation can require considerable training on partitions of the data set, with models of larger dimensionality (layers, parameters to estimate) requiring larger quantities of data and time, to avoid misleading over-estimation. Ultimately, even when modeling structures lack immediate transparency, their value toward future prediction can be immeasurable. Classification accuracy measures such as accuracy, reliability, and recall percentages can be demonstrative of fit, and the associated value of these efforts, in these cases.

When the phenomena to predict is more continuous, alternative tactics become relevant. For example, often data is interval, but comprises whole numbers with differences between values. This is common in the cases of counts (number of wins, number of failures, etc.), likely bounded theoretically by zero. In such scenarios a Poisson or binomial regression approach would be more appropriate as a model estimation approach, with Analytical Objectives captured through Chi-square statistics, for example. When outcomes are time-inter-dependent (i.e., are related serially to other observations in time), it may be appropriate to apply time-series approaches to model estimation, in the family of such techniques as ARIMA. In such instances, Box Ljung residuals, the fraction of variance accounted for, or other assessments of accuracy serve as possible Analytical Objective metrics. In highly complex scenarios, where feedback mechanisms can't be ignored, more sophisticated parameter estimation processes for system models may be required (mean square error-type estimations used as Analytical Objectives). And, of course, in cases where reinforcing or balancing loops aren't the focus, or at least not a primary concern, other methods in the realm of general linear modeling become suitable, from ordinary least squares regression in simple cases to HLM, AB-GMM, boosted decision trees, among many other options. R-square, log likelihood estimates and a host of other metrics play the part of Analytical Objectives in these cases. However, it should be emphasized that these options are suitable only if any of the previously discussed issues do not play a prominent role in the model structure.

Though even with the most accurate and representative Predictive models in place, even when we have a strong sense of "what causes what", the question of "what to do" isn't always straightforward. As discussed in

Chapter 3, Connections between Utility predictors and the predictions made, along with Objective outcomes, are often far more complex than simple linear relationships. Consider some of the methods described above, either in terms of the structure of associated models or the manner in which their performance is gauged. They include multi-level neural networks and forests, logit calculations, feedback loops, not to mention the potential inclusion of interaction terms. These are non-linear relationships. The incremental returns to increases in a single Utility will not tend to be constant. Their returns to the issue predicted will see limits and tradeoffs. Since decisions tend to be multi-dimensional, and not merely the solution of a single equation but rather a system of equations, Predictive analytical results must be considered systematically in order for them to become Prescriptive.

5.1.3 Tactics Used in Prescriptive Analysis

What is the systematic consideration of Predictive results? Consider a simple example. Let's say that our FMO is to maximize some Y, which we have predicted very successfully (high accuracy, high degree of variance accounted for) to be the result of two decisions we have control over, X1 and X2. The square of these terms also contributes significantly to the prediction of Y, although in the opposite direction (e.g., negative coefficients on these quadratic terms, where the linear terms are positive). However, X1 and X2 are investments with dollar values, and their sum is restricted to a fixed budget, X1 + X2 < Budget. We can't merely suggest increasing both X1 and X2 indefinitely. In fact, our best choices depend both on what our current state is in each of these investment decisions, as well as the level of that budget. Even before we start thinking about uncertainty in the model estimation, additional limits on X1 and X2, or other predictive factors, we're going to need a way to search for an ideal solution. That is where true Prescriptive analysis comes into play. Taking Manifested systems of description and prediction, and Explicating the best prescriptions of what to do. Consider the earlier example of the network managed by Walsingham in the effort to counter the efforts of Mary Queen of Scots. It wasn't enough to understand when encoded messages were being transmitted and to who, or to merely develop an accurate and reliable model for decoding. What yielded the critical result was a prescription of strategic timing, waiting for the most opportune moment to prescribe and execute a plan.

How do we "find" the most opportune of options? Simply stated, we search. And we don't look only for decision options that are opportune. Instead, we strive to identify globally optimal solutions superior to all other options ... tempered by a rational consideration of the costs of search. Searching itself certainly isn't the distinguishing factor that sets Prescriptive analysis apart from Predictive. When we attempt to fit models to data, we are searching as well. We're searching for structure and effect sizes. In basic OLS that search involves a simple application of calculus, working behind the scenes of linear regression estimation. In forest and system dynamics model estimation we are selecting initial numerical parameters and adjusting them (e.g., training) to accommodate increasingly better representations of the data available. In such cases, we depend on performance criteria to stop the search process. That kind of incremental adjustment, bounded by stopping criteria has a corollary in Prescriptive analysis as well.

In Prescriptive analysis, we aren't attempting to come up with model parameters (e.g., β coefficients) or model structure. We are instead explicitly leveraging an existing model structure and considering changes in other Utilities (the X variables we have a chance of influencing in practice), with an interest in pursuing superior levels of our FMO. Certainly, considering changes in those factors couldn't proceed in the absence of a model of cause and effect, thus Predictive work is fundamentally dependent on Predictive work. Similarly, however, Predictive work relies on Descriptive findings just as Predictive work does. Descriptive findings help reinforce limits on how Utilities (X decisions) can be adjusted. And these limits almost always exist, both in a univariate sense (X1 is bound by upper and/or lower limits), as well as in combinatorial terms (e.g., the sum of X1 and X2 must not exceed a budget).

If our predictive models are simple, and the landscape of cause and effect is relatively straightforward mathematically, it may be possible to derive closed-form solutions that present optimal decision points. However, the simplicity and scope of assumptions required can be rather non-representative of reality. Nice as theoretical examinations, perhaps, but typically difficult to extract direct practical value from. Often the best models that we derive from Predictive efforts are complicated, and contain elements of uncertainty that are hard to work within a closed form combined system of equations. That is, when Prescriptive efforts evolve from rigorous Prescriptive efforts, calculus might not do the trick for us. We'll have to actually look at possible outcomes associated with changes in Utilities in a step-wise search.

At the same time, we might be fortunate enough to have a relatively limited search space, allowing us to look "at everything", and simply select the best-combined option. For example, if the total number of decisions to be made, and the total number of possible values that they can take on, is fairly limited (e.g., when we have a handful of binary, or low-scope nominal decisions to make), it may be feasible to evaluation each possible combination of decisions. In each instance, the performance of combined solutions can be described by measures of expected benefit as well as risk profiles around such benefit (again, since our models of prediction include uncertainty around effect estimates as well as variance not accounted for). Depending on the Predictive model complexity, the generation of characteristic performance and risk profiles in these instances may involve simulation and the use of convergence criteria on estimation (e.g., keep simulating until estimates of risk and performance stabilize). Furthermore, the Analytical Objective driving our selection can be a joint consideration of both risk and return to each of the decision sets available. But the "search" for solutions in this case can prove to be little more than a sorting process.

When we're faced with a much more extensive set of choices, such as when there are a great number of decisions to make simultaneously, or when many decisions can take on any value within a continuous range, we likely won't have the luxury of looking "at everything". We'll tend to need an alternative approach; A search for solutions that involve stopping criteria relating to the level of FMO performance, in tandem with any stopping criteria on the estimation of such levels. When the system of equations that describe our managerial context is entirely linear (albeit potentially inclusive of noise), or if we can be convinced that the relational landscape between decisions X and outcomes Y have no more than a single multivariate mode (i.e., peak, or well), greedy hill-climbing tactics can be brought in. Uncertainty aside, you may have seen such approaches used in simple modeling settings, as made possible through the use of Excel's Solver tool. In such instances, the search-stopping criteria is simply whether or not we've reached a point where any modification to the solution would be inferior. We'll have attained the global optimum, measured by some combination of risk and return on the solutions considered. The Analytical Objective of identifying that optimum will have been met.

However, when we don't have such unimodal guarantees, or when the landscape is replete with discontinuities based on rules and thresholds put in place by design criteria or management, then we'll need to rely on search-stopping criteria that have more of a "base on what we've examined", or "based on the amount of time we've spent" flavor to it. Search-stopping criteria in simulation optimization contexts are discussed in depth, with a range of applications, in Bendoly's (2020) "Blackbelt" text. Consider for example the Traveling Salesman Problem brought up in Chapter 4. Recall the implied scope of the task of organizing a trip of even just 20 destinations. Think about how much more daunting that task would be if we scaled up to 150 destinations, as is common for drives working for package delivery services. Try calculating 150-factorial on a calculator or spreadsheet. You won't be doing a comprehensive search here. You might start with a heuristic, as in Chapter 4, and then follow up with an advanced search approach such as a Genetic Algorithm. In cases where model structures involve uncertainty, we might need convergence-stopping criteria on the simulation of risk and return to each solution as well, but the chief search-stopping criteria will need deliberate considera-tion as well. For example, should we continue to examine new, previously unexamined solutions only until we cease to see improvement in the projected average of the modeled FMO? Or until we have reached a practical time limit? Perhaps whichever is reached first? These stopping criteria, and the ability to compare the best solutions encountered at the point in which they are met provide the foundation of Analytical Objectives in such complex Prescriptive work.

Time is a resource we are constantly attempting to make the most of in these journeys. It's something we can't easily replace. Because of that, project teams and their leaders often think about analysis in terms of allocations of time from a single 'budget'. This in turn often leads to the starving of certain critical stages (often descriptive and prescriptive efforts) in place of excessive emphasis on what is often viewed to be more tangible wins (i.e., incrementally better and better predictions). A far better approach involves making deliberate efforts to move past acceptable predictions and into true solutions, as derived through prescriptive investments in time. Thomke (2001) discusses numerous virtues of successful innovative dis-covery, including the frequently touted concept of "fail early and often". As we've stated, in the big picture it's impossible to fully appreciate what a failure is unless the fit between end-users, problems, and solutions is considered. If you haven't allocated sufficient time to prescriptive analysis, your likelihood of assessing "fit" holistically, and thus identifying "failure", is going to be low. Failing early, but never closing the loop is not virtuous. Failing early but viewing failure through the lens of the full Cycle is.

5.2 Visually Augmenting Analytical Communication

Having read the last section, if you weren't already aware of it from your own related experience, you should now have an appreciation for the sheer bulk of examination needed to properly execute a soup-to-nuts, evidence-driven analytical journey. In order to guarantee a logical flow of development, there is a lot to keep track of. Using words to describe what we see, in a manner that others can understand, particularly when expertise varies across team members and hand-offs occur, can be instrumental. But words and numbers alone can also be limiting or, at least, inefficient. As effective project leaders, with an interest in ensuring that no critical details fall between the cracks in this journey, it is incumbent on us to ensure the most effective modes of communication are leveraged. These include visual ones that go beyond words and numbers alone, capitalizing on (here's that word again) structure.

The visualization of data and ideas plays a critical part in both the identification of the best problems, as well as analytical stages of estimation and evaluation. These roles are often underappreciated in comparison to the often more popularly touted role of visuals as mechanisms for conveying final solutions to stakeholders external to a project team. Critically, as a project team begins to brainstorm what their Objectives and Utilities might entail, even cursory levels of visualization can prove useful. Think about the tabular structures used in identifying and comparing candidate Objectives. Those tabular structures are in fact visual renderings, with structures that yield meaning (columns and rows are "structure" not text). Similarly, as teams move into the Connections stage, visual renderings such as RI-Fishbone diagrams allow conceptual associations across a wide range of Utilities and an Objective, and across a wide range of dimensions describing those potential relationships. Imagine attempting to put all of the information held in Figure 3.6, or Figure 4.1, into words and numbers alone.

Because of their ability to leverage structure to convey system-wide Connections, visuals allow individuals to identify patterns in the groupings, relationships, and dynamics of conceptual, as well as analytical, thoughts and findings. Consider the analogy in analytical pattern recognition. Algorithms designed to identify patterns are only as good as the inputs provided to them. The same holds for human decision-makers in project settings. We are bounded by both the availability of information, as well as our ability to process it. Strong visual structures for data provide a vehicle for making information available, but also, critically, can augment our ability to process, or make sense of, that information in a more holistic manner. Words and numbers presented in a descriptive paragraph format might provide similar content but can be constrained in their ability to augment such sense-making processes.

In a research paper discussing the effective use of visualization in research, Basole et al. (2022) outline opportunities for teams to leverage structured depictions in the process of theory/model development, in the process of theory/model testing, as well as in the process of translation/conveyance. Using the terminology of Figure 1 from Basole et al. (2022), Table 5.1 demonstrates how these applications of visualization map to the stages of the OUtCoMES Cycle.

A couple of points are worth noting with regard to this mapping. First, the majority of visual applications are focused on a singular set of stakeholders: the project team. If the team capitalizes on visual renderings of information to facilitate understanding, they will be significantly constraining their ability to communicate ideas and advance into effective analysis and solution development.

Table 5.1 The Role of Visualization across the OUtCoMES Cycle

Applications of Visualization	Role of Visualization	Primary Consumer	Challenges to Effective Use	Stages of the OUtCoMES Cycle
Theory / Model Development	Identify gaps / pose (new) model forms	Internal: Project Team	Sufficient depth and breadth	Objectives Utilities Connections
Theory / Model Testing	Estimate / validate model structures	Internal: Project Team	Sufficient metrics / assessment	Manifest Explicate
Translation / Conveyance	Dialogue / justify / assist	External Stakeholders	Critical detail clarity	Scrutinize

A second point worth emphasizing is that visualization typically provides a bridge between internal team considerations and external stakeholders (i.e., end users). In that sense, we can consider the role of both internal and external parties as findings and potential solutions are subjected to close examination (Scrutinize). Ensuring "fit", as we've discussed, is central to effective solutions. From the perspective of the project team, the focus tends to be on fit between problem structure and solution, as solutions are developed (the earlier stages having focused on fit between a problem and an end-user). Because of that focus, it is critical to ensure that the design loop is closed, with scrutiny placed on the fit between the solution and the end user. Who else is best to apply such scrutiny but end-users? However, since end-users are fundamentally using a distinct dialect in their consideration of solutions, visualizations intended to assist in clarifying the value and risks of solutions must be rendered in line with that dialect (cf. Bendoly 2016; Bendoly and Clark 2016). In short, we should expect that visuals a team uses to Scrutinize their work, are going to be substantially different from those that can best serve external stakeholders.

🜂 *What's in a Picture*

Lin works at a large software firm, critically positioned within in the healthcare industry. In her role as an analyst, Lin supports the technical account manager responsible for a large enterprise customer. Lin's manager sends a frantic e-mail reporting an escalation that has just come in from the client. The complaint was that the promised annual reductions in latency of medical record processing had not yet been delivered as stated in the firm's master service agreement.

Lin's manager asks the team to gather data showing observed latency metrics for the client over the disputed time period. The manager wants to present this in a video call with the client the following afternoon.

Lin collects the needed latency data points, enters them in a spreadsheet, and prepares a slide for their manager that provides a visualization of the data as a bar graph alongside tabulated data points.

It is immediately clear to Lin that the client is correct in the complaint. The promised improvements in latency had not been delivered to the client over the period in focus. Concerned that their manager would not be receptive to the bad news, Lin adjusted the visual representation. She truncates the Y-axis of the bar graph to make the marginal achieved reduction in latency appear with greater emphasis – ultimately, a misrepresentation.

Lin's manager fails to notice that the Y-axis on the bar graph has been truncated when hastily cutting and pasting their prepared data and the visualization into the final presentation deck. The manager shows the bar graph to the client during the video call, and points to the reduction in latency over the last six months. The client, however, is equipped with their own direct experience with the matter at hand and almost immediately calls out the misleading visual depiction. Lin's manager apologizes to the client, promising to get a diagnosis for the lack of progress, and an improvement plan to them by the start of the following week. The manager immediately messages Lin to schedule an urgent meeting between the two of them.

Learning Concept: **Information Filtering** – Omitting specific dimensions and/or records of data in analysis and presentation. While some filtering can purposeful, and appropriate, it can also be used to deliberately withhold unfavorable or conflicting evidence.

Reflection Questions

1 If you were Lin's manager, how would you handle this scheduled meeting?
2 Can you think of scenarios where information filtering would be appropriate? Would it be appropriate to ensure that presentations or handoffs of such filtering be accompanied by a statement of that action, and rationale?
3 What can Lin's manager do to make sure that poor communication between herself and her direct reports doesn't happen again in the future?

Research Follow-Up – Find out more about the forms and causes of information filtering in organizational hierarchies. Use this research to identify a checklist that you might follow in the future to avoid miscommunication related to such filtering.

The Basole et al. (2022) essay, along the Bendoly and Clark (2016), provide wealth of recommendations regarding how best to design visuals with specific use cases in mind. They also discuss the threats that both Omission and Commission (superfluous content) pose in efforts aimed at effective visual rendering. While we won't detail these methods here, we strongly recommend the reader to these sources for further reference. As we will further emphasize in the following chapter, effective communication is paramount to the effective management of team dynamics as well as the discovery and fruitful pursuit of the best problems and solutions.

Practitioner Recap

A stepwise progression through a series of analytic techniques is required to understand constraints, and to generate evidence informing an effective solution. This also ensures team members' efforts and expertise are effectively leveraged. As teams move from Descriptive to Predictive to Prescriptive Analytics, evidence gathered in one phase informs the work of the next. Connections between Utility predictors, predictions made, and associated Objective outcomes often are far more complex than simple linear relationships. Because decisions tend to be multi-dimensional and emerge not merely from the solution to a single equation but more likely from a system of equations, Predictive analytical results must be considered systematically prior to Prescriptive efforts. Prescriptive analytic models are often complicated. Having clear search stopping criteria, related to FMO performance, is appropriate when the total number of decisions and possible values are limited. In complex landscapes, with discontinuities based on rules and thresholds from design criteria or management, search stopping criteria based on examined solutions or time constraints become necessary. Appropriate time allocation is crucial in analysis, and deliberate investment in Prescriptive efforts increases the probability of true solutions. The use of visualization is not limited to the project team but can also serve as a bridge between internal considerations and external stakeholders, ensuring the fit between problem and solution. Tailoring visualizations to the dialect of end-users is crucial for clarifying the value and risks associated with any given solution.

Chapter 6

Leading Discovery in Projects and Programs

The sources of alternative perspectives, and the skills required to cover the variety of analyses that advance Objectives, seldom reside in a single individual. We have project teams for a reason. But how do we make the most of them? The following quotation sheds some light on this and helps to open our chapter discussion.

If you want to build a ship, don't drum up people to collect wood and don't assign them tasks and work, but rather teach them to long for the endless immensity of the sea.
– Antoine de Saint Exupéry, 1948 (Citadelle, paraphrased)

Teams exist within an evolving temporal landscape that is defined by shifting patterns of interactions. Teams of all kinds, across all types of organizations and tasked with a broad spectrum of goals and initiatives tend to follow a fairly well-understood series of phases of team development. This evolving pattern of interaction has implications for the ways in which discrete elements of the OUtCoMES Cycle are negotiated within teams. With tractable variation, occasionally ambiguous boundaries that blur or overlap depending on team- and team-member-specific idiosyncrasies, project parameters, and timelines, the following phases of development can be delineated: forming, storming, norming, performing, and adjourning.

6.1 The Five Phases of Team Development

6.1.1 Forming

The forming phase represents the first entry of individual team members into a team. These individuals most typically arrive onto the team from different departments within their organization, and prior or concurrent

DOI: 10.4324/9781003427650-9

memberships within other teams, operating at different phases of development. New members to teams are frequently simultaneously embedded in multiple teams, operating within different phases of team development. Thus, it is important to recognize that, as new teams form, temporally and experientially disjunctive activities and priorities can confound expectations, assumptions, and problem definitions. The influences of simultaneous membership within multi-team systems can be important as, often, the kinds of problems and challenges members are tasked with addressing across teams can carry substantial content overlaps. This can influence the ways in which current-team Objectives are internalized, understood, explained, and acted upon.

In the forming phase, members are just becoming newly oriented to the (often nascent) Objectives and responsibilities the team is tasked with addressing. Because they have also just entered into an entirely new social setting, with all of the complex dynamics that define the interpersonal interactions of newly formed collectives, members of teams in the forming phase also tend to test interpersonal, intellectual, and perspective-relevant boundaries. When members first come together, a series of questions invariably arises, including "What is my role on this team? What am I in a position to contribute to its Objectives", "Do my perspectives and assumptions coincide with the dominant mental models and framing paradigms that define this group and its orientation toward its performance Objectives? In the forming phase of team development, members begin to find points of interconnection with other members, uncovering common and divergent assumptions, ways of looking at problem solving and performance expectations. Members are likely to have social concerns orbiting issues such as getting acquainted with their new teammates, establishing the foundation for productive working relationships, figuring out normative expectations for what kinds of behaviors are acceptable – and which aren't – and learning about how other members understand their team's Objectives and associated tasks.

What also happens in the forming phase is that there often is an unrepresentative, skewed distribution in the airing of the various points of view, assumptions, and expectations. This has bearing on how the team could most profitably begin to formulate Objectives, pursue understanding and elevate performance. Some new members are likely to be reticent to vocalize their perceptions and expectations, preferring to rely on more outspoken or socially dominant others who may appear to have more familiarity with or knowledge regarding the project space. In an associated manner, the views and expectations of others who are more socially "powerful" or dominant are likely to receive more airtime as the team goes through its first series of acquaintanceship rituals. There is certainly no guarantee that the most vocal or socially aggressive new members are likely to be the most on-target members in terms of their problem focus. However, it is a common phenomenon that, at the outset, their views and perspectives are likely to have an outsized impact on the ways in which the team goes about defining its processes and Objectives. It is also likely that, given project-driven expertise convergence and a limited pool of human capital to draw on, some of the more tenured team members have prior experience working with one another. These experiences can play a role in the development of networks of communication within the new team, the development of a team culture, and the formation of goals, and practices as previous experiences color and shape how the Objectives of the new team are interpreted, communicated, and executed.

Forming in the OUtCoMES cycle: objectives and utilities: The initial stages of the Cycle, Objectives, and Utilities, correspond temporally with the forming phase of team development. In the forming phase, initial considerations of candidate FMOs and systematic exploration of Utilities begin to emerge. As noted earlier, Objectives describe the measurable outcomes that motivate project investments, provide anchorage for local reference system details, and ultimately help to define how project success is measured. Utilities, drawing from real-world considerations, describe controllable options with the potential to advance specific Objectives and encompass a range of decision variables and model parameters. The forming phase, as described above, is defined by an inherently nascent/ exploratory character, where members' roles, understandings of one another, crucial aspects of the project, key definitions, and down-stream targets all are emergent and in flux. Nothing concrete is established at this phase of team development, and there is wide latitude for the emergent unfolding of generally agreed-upon approaches, rules of engagement, premises, and boundaries. In short, this offers fertile ground for the documentation of alternatives. As described in earlier chapters, premature rigidity in the development of fixed trajectories and assumptions about various aspects of the team's project can lead to poorly specified/under-specified parameters that ultimately generate significant wasted time, non-trivial resource losses, and unmet stakeholder expectations. Effective navigation of the forming phase is critical as it allows for the maintenance of fundamentally exploratory ideation orbiting understanding of both the team's candidate Objectives and Utilities.

6.1.2 Storming

Once the configuration of the team has coalesced around its principal membership, with the odd member coming and going throughout the team's development depending on shifts in project parameters or broader organizational needs, the team enters the storming phase. This phase typically involves a high degree of interpersonal friction, emotionality, disharmony, and even hostility between members of the team. The storming phase is often the most difficult period of team development to emerge from successfully – with potentially wide-ranging, non-trivial consequences for failure.

Tensions that emerge between members during this phase, potentially contributing to the risk of suboptimal performance, can arise from a number of related sources. These can include, for example, personal issues that members themselves experience as a consequence of being on a new team, in a new situation, and faced with new challenges and uncertainties. Personal issues can impact a whole host of processes within the team, from how members interpret their own responsibilities within the team, to their commitment to the team's performance goals, to how well they are able to navigate the interpersonal complexities that define novel social situations. By extension, these drivers of tension, friction, and hostility also can include a range of interpersonal issues – not the least of which are cleavages that surface along professional fault lines. Although there is likely to be a great deal of overlap in the ways in which the members of new project teams tend to understand the team's performance spaces, its technical aims, and the technologies that define its operational trajectory, this understanding also is likely to be defined by a great deal of interpretational idiosyncrasy. People – members of project teams or not – don't always see eye-to-eye. Professional discrepancies can generate heated, even hostile, inter- personal dynamics. Such clashes can lead to prolonged periods of interpersonal tension that substantively impede any kind of progress being made toward the team's immediate, intermediate, and even longer-term Objectives.

It is not uncommon to see these periods of potentially prolonged infighting defined by the emergence of shifting coalitions of interests, perspectives, and stakeholder aims, or even social cliques that coalesce around dominant personalities and politically central individuals within the team. These emergent – but often transient – subgroups may form around expertise fault lines defined by occupational anchors, areas of emergent agreement (or disagreement) about definitions, priorities, ap- proaches, or project expectations. Conflicts also may develop as members compete with one another in a political sense to impose their assumptions and preferences on other members of the team, and in so doing play an influential role in the shape and trajectory of emergent project parameters and dynamics. All of these processes play a role in how the team ultimately comes to define itself, as well as the Objectives it is responsible for accomplishing.

What also typically emerges in the storming phase are important changes related to members' understanding of one another and, in an associated sense, one another's priorities, perspectives, and mental models (in broad strokes, at least). Members' project and task agendas are slowly revealed to other members, who can begin to develop some clarity as a consequence of on-going, and very frequently heated, discourse and conversation about what the team is tasked with accomplishing, and also how it will go about accomplishing it. It is also in the storming phase that the members of newly formed teams slowly begin to develop some understanding of one another's work and interpersonal styles. As the storming phase begins to mature, members start to turn their attention toward both obvious and more tacit performance obstacles. Addressing these roadblocks is critical, as they have the potential to impede or even entirely short-circuit the advancement of FMOs. As the storming phase starts to come to a close, members begin to increase their efforts to identify ways both to generate functional solutions in the service of the team's Objectives, but also actively pursue avenues that can help to meet individual team members' needs. Alternative courses of action (and Objective and Utilities for focus) can be further delineated to that end.

As implied by this discussion, the storming phase is part of a "critical zone" in team development, with substantive long-term implications for how well the team is able ultimately to navigate its official charge. Whereas successes experienced by the team collectively during the storming phase can help to lay a strong foundation for the achievement of long-term performance gains, failures that the team experiences during storming can plant the seeds for long-lasting disfunction that can ultimately derail any chances for achieving collective success. It is critical that the storming phase be negotiated successfully so that the team enters the next phase with an eye toward generating collectively achieved measurables.

Storming and re-calibration within the OUtCoMES cycle: The initial stages of the Cycle also correspond with the storming phase of team development. In a definitional sense, the storming phase is characterized by a high level of interpersonal and professional turbulence. It is also identified by an ongoing evaluation and reevaluation of a host of strongly held assumptions and expectations as members struggle against one another, and one another's ideas. These struggles, though daunting, have an important role to play. They generate operational and interpersonal clarity bearing on the identity of the team, its Objectives, starting points, and end goals. Critical to successful project execution is integrated architectural dynamism that systematically impedes premature solidification of key project definitions. In the storming phase, teams engage in on-going reconsideration of initially agreed-upon or initially supported candidate FMOs and derivative Utilities of focus. The turbulence of the storming phase provides that Objectives and Utilities remain explicitly subject to on-going examination, recalibration, and negotiation. In this process still other alternatives germinate. This turbulence serves the functional goal of allowing both of these sets of values and ideas to remain dynamic, pursuant to the initial development of Connections between Objectives and Utilities.

During this dynamic phase of the team's development, examination of underlying assumptions and predicates can be extremely productive, as there has been no opportunity for substantial inertia to have developed. The turbulence of the storming phase is challenging but virtuous. It affords teams the space and opportunity to pull back from ill-conceived, incompletely or insufficiently developed, or erroneously anchored premises which could otherwise represent a fatal flaw in the foundations of the project.

6.1.3 Norming

At the outset of the norming phase, team members are likely to have generated at least a tentatively agreed-upon understanding of the nature of the project, a well-vetted FMO and corresponding analytical Objective, a list of high-value Utilities and alternatives, and a rough sense of the system of Connections at play. With analytical Objectives outlined, members also will likely have some preliminary ideas regarding approaches to working through unknowns, to developing tentative definitions for evolving/moving parts of the project, and to a process for reconciling emergent deviations from previously held assumptions about process, procedures, and the FMO. It is in the norming phase that members begin a process of active, and ostensibly productive, cooperation with one another. They begin working through complex problem solving by building and relying on one another's expertise and experience. Members begin to develop and codify rules bearing on acceptable conduct, processes for addressing grievances, and approaches for navigating intransigent definitional and interpersonal issues that inevitably emerge among groups of motivated professionals with their own assumptions, coding classifications, aims, and interests.

The architecture of a natural, organically emerging hierarchy also begins to take shape. The emergent organizational structure often takes on an entirely different character and dimensionality than would otherwise have been expected were one to have viewed a snapshot from the forming phase. If the storming phase has been traversed effectively, the emergent sense of leadership that manifests within the team will reflect more expertise/experience/knowledge-based anchorage than it will political/social/coalitional anchorage. Members equipped to lead – defined operationally – begin to occupy and ultimately fulfill key roles within the team's configuration of interconnected elements of the process. In the norming phase, the interpersonal hostilities that defined the storming phase begin to fade into the background and ultimately diminish, as a focus on maintaining interpersonal harmony and collective momentum is emphasized. Here, however, divergent viewpoints may still be discouraged, which can have a consequential dampening effect on important objections which could otherwise have helped the team from falling prey to group think, and time and energy devoted to otherwise avoidable dead-ends and project traps.

As with the storming phase, the norming phase also is part of the critical zone of team development, carrying with it important downstream implications bearing on the team's ability to achieve its Objectives. When the norming phase is navigated effectively, members can develop growing feelings of affiliation for one another, and for the team as a whole, identifying with both teammates and the team alike. Successful norming also can help the team to develop a coherent division of tasks, duties, and responsibilities. It can also result in shared expectations bearing on both how work can and will be accomplished, as well as understood parameters for what successful team outcomes are likely to look like. Successful navigation of the norming phase also can serve as an experiential buffer against the disintegration of the team when it faces inevitable setbacks, short-falls, disappointments, and failures along the way. Successful norming provides a context for members of the team to look to one another for inspiration and support when riding through these turbulent waters. Within teams where this norming phase has been worked through successfully, it can often feel like holding the team together becomes more important than actually accomplishing the team's formal Objectives. Whereas successful norming can help to provide teams with substantial interpersonal and social resources to bolster their persistence and perseverance, failure during norming can leave teams vulnerable to unavoidable turbulence and diminish their chances to ultimately accomplish their performance goals. It is essential that teams work through the norming phase well so that they enter the performing phase equipped to execute their tasks effectively.

Norming and the development of connections: The Connections stage of the Cycle corresponds temporally with the norming phase of team development. In the norming phase, identification of relationships between Objectives and Utilities begins to emerge and to take on recognizable dimensions that are likely to be shared across members of the team. As noted earlier, Connections reflect the relationships between Objectives and Utilities, capturing either cause-and-effect or correlational dynamics, encompassing a potentially broad scope of constraints. Connections can be defined by a range of characteristics determined by understood distributions, and can involve a range of relationships and feedback mechanics – all of which are subject to evolving interpersonal and social dynamics present in the norming phase of team development. The norming phase, as described above, is defined by ongoing evaluation and reevaluation of assumptions and expectations as the team works toward clarification of central facets of the project, and underlying assumptions driving understood points of departure. Connections based on preliminarily agreed-upon Objectives and Utilities begin to take on a formalized structure in the norming phase – to the extent that this level of confirmation is possible. These agreed-upon depictions, limitations, and a portrait of the realities defining the project emerge during the norming phase. This common framing ultimately helps to define the operational space in which the project is embedded, creating ideal conditions for the Manifest stage of the Cycle.

6.1.4 Performing

As the anchor label suggests, teams in the performing phase have typically reached a higher level of maturity – a sufficient level of maturity – to effectively execute collective tasks, duties, and responsibilities in ways that don't generate hostilities. This capacity is typically associated with a higher level of structural and procedural organization, and less frequent intra-team interpersonal and social turbulence or chaos. The operation and functioning of teams in the performing phase will tend to be relatively smooth. Performing is the phase of team development where members are able to begin to systematically integrate their disparate domains of expertise within their project sub-groups and are able to work through complex problems and tasks in creative ways without instigating interpersonal tensions and conflicts. In the performing phase, teams tend to operate with a relatively clear and stable structure that is broadly agreed-upon and understood. Members are motivated more by the accomplishment of the team's project Objectives than by personal political or social momentum. The primary challenges teams face in the performing phase are continuing reflection on how the team operates, continuing reevaluation of the underlying assumptions made about what the team is working to accomplish and how it is going about accomplishing it. In this phase, the team must also continue to build and maintain functional and productive working relationships that allow everyone associated to effectively integrate their collective efforts in the service of creative, problem-focused work.

   Performing and the explicate and scrutinize stages of the cycle: The final three stages of the OUtCoMES Cycle, Manifest, Explicate, and Scrutinize, correspond temporally with the performing phase of team development. In

Table 6.1 Team Development Phases and OUtCoMES Cycle Stages

Phases of Team Development	Stages of the OUtCoMES Cycle
Forming / Norming	Objectives
	Utilities
Storming	Connections
Performing	Manifest
	Explicate
	Scrutinize

the performing phase, members have put aside interpersonal frictions that can impede team members' execution of core tasks and activities. A healthy dynamic of reevaluation and recalibration of maintained assumptions and definitions is now in place. In the performing phase, members' integration of their experience and expertise allows them to dig deeply and creatively into the complex dynamics that define the team's project. The team is now in flow and makes systematic progress toward project Objectives.

As noted earlier, the Manifest phase of the Cycle encompasses the assembly and codification of the structure of the system of Connections relating to identified Utilities and Objectives. It creates a foundation for a search for problem and context-relevant solutions; while the Explicate phase encompasses the generation of both predictive and prescriptive Utility parameterizations to address the operational aims of the project. The actionable intelligence that emerges from the Explicate stage of the Cycle is developed based on the results from the previous stages. In the Explicate and Scrutinize stages, project results are evaluated, Objectives and Utilities are modified, and new Connections are developed based on the proximity of results to agreed-upon Objectives. The common frames of reference that emerge here operationally serve as a foundation for the execution of tasks that constitute the primary focus of activity during the performing phase. The mapping of this and other team development phases to the Cycle is summarized in Table 6.1.

6.1.5 Adjourning

Once the team's project has been completed and as the team prepares to deliver to key stakeholders, it enters the final phase of team development. In the adjourning phase, members prepare to achieve closure associated with both the common work of the team, and the social bonds that have developed during this performance period, and ultimately, to disband. Because organizations are bounded ecosystems, among the critical aspects of a successful adjournment is that members are positioned to successfully move into future teams, likely with at least some overlapping members in common with their previous team. It is important that project teams disband with the authentic sense that important perform-ance Objectives have been accomplished. Adjournment can be an extremely emotional period for everyone involved. Team members, who have worked together intensely for an extended period, will have developed common bonds forged as a consequence of having gone through periods of turbulence, creativity, productivity, accomplishment, achievement, setbacks, resurgences, and ultimately completion of long sought-after aims.

In the adjourning phase, it is important to acknowledge the specific contributions made by everyone involved in the project, praise their efforts, and celebrate the successes of the team. Because team memberships will tend to overlap over time, and members are likely to find themselves repeatedly working with the same individuals across multiple projects, teams should ideally disband with members left with the sense that they would like to work with these individuals again in the future. Successful adjournment follows a specified number of interactions, following which the team's formal activities are suspended, either temporarily or permanently – as the results from the team's efforts are evaluated by key stakeholders, and conclusions drawn as to the next steps for the project and the team.

🌍 *Breaking to Fix*

Salma is responsible for a group of global teams. Two of these teams, one located in Hyderabad and the other in New York, are collectively responsible for the same customer outcomes. Ambiguity in the sharing of responsibilities, however, has eroded the sense of accountability for failures in their intertwined software systems.

Poor delineation of concerns (and causes), paired with this lack of defined accountability, makes it difficult to resolve what has become a pattern of repeated failures in the software. Salma finds that the teams keep passing blame back and forth, instead of working together effectively to diagnose the underlying issue. Each team is becoming increasingly defensive, and reluctant to share information on their respective systems with the other team. There is a sense that the other team could weaponize such potentially compromising information in assigning blame. Instead of collabo-rating, each team continues pulling back more and more.

Salma ultimately decides that the situation is intolerable due to the ongoing impact on customers. She assigns the problem to the Hyderabad team and gives the New York team an entirely different set of different responsibilities. This places ownership of the entire software system unambiguously on a single team in one location. The approach works well, as the Hyderabad team takes holistic responsibility for looking across the entire software system, and is then able to identify, mitigate and ultimately resolve the root cause of the repeating failures. Salma is left with a resentful New York team, and shifts gears to begin addressing the challenge of rallying them around their newly assigned charter.

Learning Concept: **Superordinate Goals** – Overarching goals whose accomplishment depends on the combined achievements of two or more parties.

Reflection Questions

1   Was Salma's solution to this inter-team failure the only possible way to resolve this situation?
2   What role could superordinate goals play in changing the outcomes of this situation?
3   If Salma thinks long-term as well as short-term, would she have tried to resolve this situation differently?

Research Follow-Up – Find out more about superordinate goals and their role in team dynamics and conflict management. If you end up in a situation like Salma's someday, how might you use this knowledge to good advantage?

6.2 The Role of Project Leaders

The complex architecture of most projects creates subdomains that are headed by various team members with area-specific experience and expertise. However, project leaders or project managers are responsible for the trajectory of the team's work, shaping the team's culture as a whole and facilitating the team's progress toward project Objectives. These individuals can have a significant influence on a team's ability to successfully negotiate the phases of team development described above. There are several specific ways in which project leaders can help to advance the emergence of a productive culture of performance within a team.

6.2.1 Focusing on Conflict

Through the forming phase, when project teams are starting to evaluate Objectives and Utilities, conflict interactions inevitably emerge between members of a team. This is particularly common in newly formed teams composed of highly motivated professionals with well-developed professional identities and a long list of technical expertise. Two types of conflict generally tend to pervade project teams: task conflicts and relationship conflicts. While task conflicts typically involve functional disagreements relating to task-relevant issues such as project parameters, understood mechanical associations, and underlying problem dynamics, relationship conflicts generally are emotionally intense interactions reflecting incompatibilities relating to members' values and personal issues. While task conflicts can help project teams to uncover hidden underlying weaknesses in the definitions attached to key variable relationships, sources of bias in operational definitions, or inconsistencies in the ways in which project parameters have been codified for analysis, relationship conflicts are likely to systematically impede the generation of any novel or relevant insight bearing on critical elements of the project.

Rather than serving as a forum for digging deeper into important aspects of a project, relationship conflicts tend to lead to feelings of animosity, mutual dislike, and distrust that can prevent members from putting their heads together in functional ways to tease apart underlying project dynamics. Project leaders focused on advancing project Objectives in the forming phase can work to shape conflicts, serving as a referee and adjudicator for task conflicts to ensure that these interactions are navigated judiciously and developed with a systematic functional emphasis on task-relevant topics. A project leader can serve as a counselor and advisor to help navigate the emotionally laden waters stirred up by relationship conflicts that can undermine the development of important personal relationships. Encouraging members in conflict to focus on the problem, and not the person, is an important step in the service of achieving project success.

6.2.2 Focusing on Knowledge

What also is likely to be of value during the forming phase, where the emphasis is on trying to pin down Objectives and Utilities, are efforts devoted to the facilitation of team knowledge assimilation. Knowledge assimilation assumes the presence of preexisting mental models or schemata which define various aspects of a problem space. New members bring individual-level mental models with them to the team, anchored in large part by domain specific expertise and experience, and it is ultimately through an iterative series of conflicts and negotiations, and lobbying sessions that these individual-level schemata emerge into a collective mental architecture that reflects various aspects of the mental models of the individuals who contributed to it. These team-level schemata are likely to begin to emerge, and even to coalesce in some substantive ways, during the forming and through the storming and norming phases of team development. During these phases, the members of project teams repeatedly engage in intense interactions that expose personal and professional positions, preferences, and assumptions bearing directly on key aspects of the project, emphasizing their own frames of reference and underlying domain-centric biases and values. It is in these phases of the development of the team that the mental models of the team as a whole begin to take identifiable form. In order to be of the most direct benefit to the end goals of the project, these collective mental structures must be as representative as possible of the full spectrum of insights and positions and expertise available from within the team, capturing the architecture of the most relevant aspects of the domains in which members are embedded. It is incumbent on the project leader to actively encourage and facilitate an ultimate collective structural architecture that reflects the broadest possible cross-section of member positions.

In the storming and norming phases project leaders can continue to emphasize members' focus on the problem and not the person, as well as the assimilation of knowledge to reflect as broad an integration of members' positions and perspectives as it is possible to generate. What also is likely to be of value during these phases, where the emphasis remains on refining and reevaluating, and testing assumptions bearing on Objectives and Utilities, but also in mapping out likely Connections, are efforts to facilitate team knowledge transformation. While team knowledge assimilation reflects the integration of new knowledge with existing knowledge structures, knowledge transformation reflects the development and refinement of routines that facilitate combining new and existing knowledge for actual use toward accomplishment of project Objectives. It is essential to maintain this mindset in order to maximize the collection and assimilation of the most practically leverageable knowledge – and not just knowledge for its own sake.

6.2.3 Focusing on Communication

In the norming phase, the Cycle's emphasis is squarely on developing insight into Connections that codify relations between Objectives and Utilities, patterns of member conflict have been largely addressed, and momentum driving knowledge assimilation and knowledge transformation has been set into motion. At this point, leaders can help facilitate generation of expedited, syllogistic insight into these critical relational dynamics with a focus on helping project members increase the efficiency and effectiveness of the patterns of communications within the team. In complex project settings, leaders can drive team communications using a number of different approaches. Various communication strategies have been identified as being effective in complex performance contexts, such as those embedded in project spaces in which the Cycle is likely to be most relevant for accomplishing functional outcomes.

Among the most effective are implicit communication strategies emphasizing performance goals, which reflect deliberative member communications, in contrast with, for example, approaches reflecting aspects of the setting or context, which reflect reactive communications. During the norming phase, leaders can take advantage of coalescing patterns of member interactions by systematically embracing, endorsing, and supporting a proactive emphasis on communications encompassing an explicit focus on the team's project goals – similar in an analogous sense with an emphasis on task versus relationship conflicts. Deliberative member communications, which emphasize project goals, can serve as an anchor which can help to focus members' attention on the key moving parts of the project. They can also increase the efficiency with which members work through the complex relational dynamics that define associations between project Objectives and Utilities.

6.2.4 Focusing on Learning

In the performing phase of team development, while the team is occupied with the systematic integration of members' expertise toward the execution of series of complex, interrelated, and interdependent tasks, the Cycle emphasis is on generating actionable intelligence reflecting predictive or prescriptive relationships (i.e., Manifest and Explicate), potentially involving a range of heuristics and mathematical approaches, and also on thorough evaluation of the outcomes of previous stages of the Cycle (i.e., Scrutinize). In this, likely the most consequential phase of the team's development, project leaders can help generate the kind of attentive, questioning, investigative, exploratory team culture necessary for success in these phases of the Cycle with a focus on team learning.

Team learning is an ongoing, active process. This process orbits both contemplative and critical reflection. The former is crucial for success in the Manifest and Explicate stages of the Cycle, where functional insights are being pursued. Critical reflection is essential for success during the Scrutinize phase of the Cycle where pragmatic evaluation informs the team's future focus (and re-focus) and ultimate trajectory. Team learning is characterized by process focused questioning, deliberative feedback seeking, systematic experimentation, and problem-focused reflection emphasizing the coherence of observed results. It is also characterized by on-going discussion of deviations from predictive models and unexpected – or poorly understood – process outcomes. The team learning process, and its generative outcomes, provides an experiential and intellectual basis for success in the Manifest, Explicate, and Scrutinize stages of the Cycle because it can afford teams with creative, data anchored insights bearing directly on the team's goals.

6.2.5 Fostering Psychological Safety

The value of the OUtCoMES Cycle as a logical decision-making vehicle depends entirely on the materiality and authenticity of the full range of inputs entered into the system by project team members and leaders. It also depends entirely on the openness of the team, and the team's leadership, to these authentic, materially critical inputs. Importantly, this includes openness regardless of (and, likely, perhaps in spite of) non-trivial deviations of these inputs – and the conclusions they imply – from other's expectations. Deviations from expectation can emerge from previously held assumptions, politically or socially anchored ideation, resource-embedded momentum, personal preferences, convenient narratives, or any other unsupportable – but nonetheless strongly held – positions of key stakeholders with the power to drive the team's trajectory and momentum; whether these are members of the team itself, the team's leadership, or external gatekeepers. The iterative analysis core to the S-A3 documentation represents an archetypal higher-order cybernetic system. The sequential mechanical transactions between the system's inputs and its subsequent, downstream (data-driven) recalibrations require an operational context where those inputs, and inevitably non-neutral reactions to these inputs, flow through relatively socially and politically unimpeded channels of communication, discussion, and analysis. This kind of systemic transparency is only possible in social contexts distinguished by psychological safety.

Within project teams, psychological safety reflects key stakeholders' assessment that the environment in which the team operates is safe for personal and professional risk taking. Psychological safety can play a key role in members' willingness, for example, to share data that depart from expected parameters, voice concerns with analyses perceived to be defined or impacted by underlying inconsistencies, seek feedback relating to key assumptions or interpretations, disclose errors, ask for help, experiment or go out on a limb to propose or develop creative ideas and solutions. The validity of the S-A3 as a documentation vehicle for the Cycle depends on the authentic inputs and transformations of members in the way that a performance vehicle depends on high-octane gasoline to reach its maximum speeds. Dilute the gas and clog the fuel injection, and the motor doesn't get the volume of enriched fuel it needs to propel the vehicle. When leaders work to generate a team culture defined by psychological safety, the space provided by this buffer systematically enriches the fuel entering the system, and flushes the fuel injection moving the fuel in an unimpeded way into the combustion chambers, laying the groundwork for high performance.

6.2.6 Collective Knowledge

One of the common features linking teams of individuals collectively engaged in complex work is accumulated knowledge and information leveraged in the service of the team's performance Objectives. Project teams of all kinds amass, generate, combine and ultimately leverage a great deal of project relevant knowledge and information over the course their project's development. This knowledge and information take on a usable form collectively, available to project members as needed in the execution of their tasks, duties, and responsibilities following its consolidation and stabilization, which facilitates its integration into the team's long-term collective memory. This happens as a consequence of three conceptually distinct, but related sequential processes labeled information encoding, storage, and retrieval.

During encoding new project-relevant knowledge and information are drawn into the project from various sources both internal and external to the team, where it is systematically transformed into useable mental representations coinciding with key project anchors and associated with existing knowledge in the team's collective memory to create memory traces. Memory consolidation continues through the storage phase of this process. Storage, defined as maintaining knowledge and information over time, enhances the memory traces' accessibility by reorganizing and integrating it with existing knowledge. Finally, retrieval, which is defined as accessing stored knowledge and information, involves recovering memory traces strengthened during storage.

The ways in which information is encoded, stored, and retrieved by the members of the project team can have a fundamental impact on the S-A3 process. And as shared or common vernacular is used to create project specific nuance, this both fosters efficiencies in the recovery and use of shared knowledge and information, but also increases its value by imbuing it with project-specific cadence, meaning, and weight. The idiosyncrasies in meaning, representation, and dialectic emerge over time as project teams develop filtering mechanisms that channel only the most relevant or useful information into the team's stored frames of reference, simultaneously filtering out irrelevant or extraneous information. This increases the potency of the iterative exchanges that define the S-A3 by imbuing it with both greater precision of meaning and emphasis, clearing out distracting project-irrelevant anchors from consideration.

6.3 Project Team Member Roles

Finally, among the key defining attributes that distinguishes successful from unsuccessful teams across a wide range of performance contexts and domains of work, including project teams responsible for the execution of complex engineering and design projects, is that members tend to adopt different roles within the team that satisfy very distinct kinds of team-level needs – professional and interpersonal. Effective enactment of these roles is critical if members are to have any real chance of successfully coordinating their efforts, integrating their disparate domains of knowledge and expertise, iteratively evaluating and reevaluating underlying assumptions and relationships, developing novel and creative solutions to intransigent issues, being proactive and problem focused. These roles fall broadly within the domains of task activities and maintenance activities. Members who devote regular and sustained efforts to the accomplishment of task activities contribute directly to the mechanical achievement of the team's project Objectives. As an underlying expectation, the team is composed of individual specialists in disparate domains of expertise who are on the team for a specific purpose, and who are expected to contribute to the team's project goals in ways that coincide with their expertise.

More than likely, the majority – most – members of project teams assume on-going responsibility for task activities, and so actively effectuate task roles on their team. Some, however, also contribute in other ways beyond their specific domain expertise. These individuals may, in fact, relegate some of their task-specific responsibilities to members with adjacent competencies in pursuit of these maintenance aims. Thus, in contrast with the task-focused activities of the majority of the members within the team, maintenance activities reinforce and facilitate the emotional aspects of the life of the team as a highly interdependent, highly dynamic, multifaceted social system. Members adopting responsibility for maintenance activities take on a host of important responsibilities. They can help to facilitate regular functional, communications between team members, and also between member factions in active disagreement with one another. As subject matter experts, they can help to promote productive collaborations between team members with complimentary domains of expertise. They can help to promote healthy (and simultaneously help to prevent unhealthy) forms of intermember conflict. What they can also do is serve in a liaison capacity to help strengthen lines of communication between the project leader and the team to maintain an active and on-going productive discourse that facilitates project momentum and helps the team to steer clear of avoidable, downstream obstructions.

Highly functional project teams demonstrate what is often referred to as distributed leadership. In teams with distributed leadership, most – if not all – members regularly actively engage in task behaviors that serve the team's performance aims, but also regularly contribute to the team's social health by regularly engaging in maintenance behaviors as well.

Practitioner Recap

Considering team dynamics and understanding their impact on performance should remain in sharp focus throughout The Cycle. During the forming phase, team members learn about one another and the project, and develop relationships with their teammates and the work. This exploratory relationship dynamic aligns well with The Cycle work of ideation, understanding, and alternative creation in the Objectives and Utilities stages. In the storming phase, team members begin to better understand one another's priorities and perspectives. The natural frictions that emerge, and the risks these dynamics generate, can be mitigated with ongoing consideration of alternatives that meet both project and individual team members' needs. If the challenges of the storming phase are successfully navigated, the Connections reflecting relationships between Objectives and Utilities are likely to be shared across team members as they enter the norming phase. A clearer picture of the project, limitations, and assumptions emerges, and can help to define the operational space within which the project operates, creating ideal conditions for the Manifest stage. This alignment allows the team to develop a deep understanding of the complex system dynamics as they execute the Manifest, Explicate, and Scrutinize stages of the Cycle during the performing phase. Ensuring a successful adjourning phase creates an environment for success if, or more likely when, these team members find themselves working together again.

Chapter 7

Organizational Memory and Learning

Leadership of projects clearly has implications for the effectiveness of both the discovery of the best Objectives as well as the success of their pursuit. But what makes a good project leader. Consider the following quote, as relevant today as it was when it was first articulated.

> *A leader is best when people barely know he exists, not so good when people obey and acclaim him, worse when they despise him. But of a good leader who talks little when his work is done, his aim fulfilled, they will say, 'we did it ourselves'.*
> *– Attributed to Lao-Tzu, Tao Te Ching, 5th-6th Century BC*

In the previous chapter, our discussion around project team management made multiple references to the indispensable role of project leaders. However, while outlining actions great leaders take to develop and maintain a team in its journey through the OUtCoMES Cycle, we never did provide an operating definition for "leadership". To be clear, leadership, as a topic, has undergone systematic consideration by scholars across a range of disciplines, for thousands of years (as the above quotation provides testimony). Leadership has been of intense and sustained interest to on-the-front-lines practitioners from the military to law enforcement, from medicine to agriculture, education to for-profit organizations, from the hard sciences to politics. It is the subject of thousands of published academic and popular press books, with, conservatively, hundreds of thousands of published academic research articles touching in some way on its role.

7.1 Project Leadership at Its Best

Leadership is among the most, if not <u>the</u> most, widely and deeply examined phenomena in the organizational and social sciences. Not surprisingly, in

DOI: 10.4324/9781003427650-10

light of the tremendous interest it has generated, and its broadly ubiquitous relevance, within such a wide range of performance settings and contexts, leadership has been defined conceptually and operationally by academics and practitioners alike in scores of different ways over the past decades. The current discussion is not the appropriate venue for an in-depth treatment of the full spectrum of leadership theories, forms, and dimensions. However, in light of the dynamic, technical, reflective, iterative, collective attributes of the analytics and engineering projects encompassed and informed by the OUtCoMES Cycle, there are a number of related dimensions of leadership which have clear relevance for the effective execution of complex project work.

It is likely the case that engineering and analytics project teams can benefit from the exercise of a number of different types of leadership. However, one form of leadership in particular is likely to yield the most direct, universal benefits across team development phases and stages of the Cycle. This form is referred to variously in the academy as *shared leadership*, *empowering leadership*, and/or *collective leadership*. We will delve into each of these in turn, as they all share the common feature of systematically reorienting responsibility for advancing project Objectives, formulating Utilities and Connections, and engaging in the activities of Manifest, Explicate, and Scrutinize. This reorientation fundamentally shifts responsibility for generating successes in these stages away from the exclusive purview of the project manager as an individual, and firmly into the hands of the members of the project team as a group of individuals who share collective accountability for the team's success.

7.1.1 Shared Leadership

It would be hard to imagine a single leader who was personally in full possession of the relevant knowledge, skills, abilities, and other character-istics and attributes necessary to handle all of the disparate aspects of contemporary multi-phase, multi-domain, multi-disciplined knowledge work. Yet, traditional theories of leadership have tended to emphasize the downward influence exerted by the leaders on followers, their activities, their interactions with one another, the mental models they use in the depiction and execution of their work, and in defining what project success can and should look like. That is, in traditional models of downward-focused leadership, the reality of the project – the way the project and its interconnecting facets are seen, described, and executed – is essentially defined by one individual.

Such ostensibly one-directional, downward influence has the potential to impact more than the architectural realities that define the project. It can also have a significant impact on the ways in which followers (intentionally used term) understand their organizational roles, their responsibilities vis a vis the project, their approach in the execution of their tasks, the cadence of project activities, their sense of autonomy and flexibility in defining key milestones in the evolution of the project. It can also impact how personally responsible they feel for the achievement of key project Objectives, their ownership of the project, and their sense of identification with the team. Importantly, such unidirectional influence also has a significant and consistent impact on the ways in which members, normatively, understand the nature of the project in terms of how it should be executed. Critical and creative voices of dissent are typically muted when leadership flows downward because members' sense of responsibility for this important aspect of complex work, where operational ends are in-flux, and procedural mechanics are as yet defined, falls onto the shoulders of the leader and not themselves.

The changing reality of organizations as complex interconnected systems, and the changing nature of the knowledge work organizations are increasingly tasked with accomplishing, informs a growing focus on what has been referred to broadly as *sharedness*. Sharedness reflects a distribution of responsibilities across the members of collectives, such as project teams, whether members hold formal authority to define/instigate/modify/project-critical work and activities or not. This flattened organiza-tional (read project team) architecture increases the breadth, and direc-tionality, of knowledge and information flow, both between members of the team itself, who feel a responsibility to share thoughts, insights, concerns, and questions; but also between members of the team and team leadership. Information, insight, and influence both flow horizontally across members of the team, but also flow vertically between followers/team members and the team's leader. This directionality represents a fundamental departure from traditional downward-flowing leadership approaches.

One of the implications of sharedness, beyond representing a broad shift in the culture of the team pertaining to leadership, is that relevant insights flow upwardly as well. This mechanically broadens the sources of influence within the team, helping to shape and define (and, importantly, re-define) the frames of reference and approaches adopted by the team. It encourages a reconsideration of alternatives documented in structures such as the S-A3. When leadership is shared, members experience and exert mutual influence with other members of the team. Members also experience mutual responsibility with other members of the team. The presence of these mutually influential member relations facilitates cooperative cogni-tions bearing on the structure and parameters defining the project, and the iterative, investigative, and explorative cadence of the unfolding stages of the Cycle over time.

In complex, iterative, nuanced knowledge work, unidirectional, downward-focused leadership can systematically impede, even preclude, the development of the kind of sensitive, explorative, reflective, iterative orientation critical to the success of complex analytics and engineering projects. If the mindset of the members of the team is that authority and responsibility for project parameters and dynamics fall outside of their scope of involvement and accountability, this can set the stage for very dangerous linear thinking that keeps them from exploring tangents, examining underlying assumptions, and asking critical and timely questions; i.e., it impairs the ability to Scrutinize in general.

When members can't rely on their own creativity, ingenuity, experience, and insight, but feel penned in by the formal parameters determined by distal leadership influences, the foundations underlying the architecture of a complex project can become potentially very fragile, with an increased likelihood of landing off-the mark at project's end (read: lack of fit between end-users, problems and solutions). If members normatively adopt or accept a position simply because the leader stipulates that position or parameter estimate or downstream conclusion, the implication must be that members have a responsibility to accept this position, They are unlikely to feel empowered or motivated to go through the laborious, iterative intellectual work necessary to instantiate the S-A3 in an effective way that coincides with a more/most-likely-to-be successful set of alternatives.

In contrast, shared leadership sets the stage for project members to take ownership and responsibility for a range of project variables. In so doing, it also avoids the premature cessation of the critical thinking – and action – necessary to effectuate productive downstream solutions.

7.1.2 Empowering Leadership

While shared leadership reflects integrated responsibility and accountability that extends beyond the team's formal leadership group, and through to the project team's membership, empowering leadership further encompasses the idea of a heightened level of autonomy in the pursuit of key project Objectives. Empowering leadership carries with it the specific aim of increasing members' motivation to work through creative and intellectual obstacles and to achieve collective outcomes. This can include motivation to identify an FMO and associated set of alternative candidate Objectives, working creatively to identify Utilities, thinking outside of the box to flesh out obscure or not-immediately obvious Connections.

Empowering leadership relies on a variety of leader behaviors and actions. These can include leading by example, participative decision making, or coaching. Empowering leadership also can include behaviors such as fostering participation in decision making (e.g., delegation), and expressing confidence in team members' performance. Behaviors such as leader coaching, where the project's leader provides feedback and guidance intended to increase the effectiveness and efficiency of members' execution of project tasks and activities and situates team members to be successful by incorporating the lessons from coaching into their project-focused repertoire. What coaching also does is set the stage for members to execute these tasks and activities autonomously, beyond the direct linear influence of the leader.

Leader behaviors, such as delegation, systematically increase the range of activities over which the team's members have responsibility. Delegation also offers a relatively well-defined space in which to allow team members to learn, grow, experiment, and receive feedback about key aspects of the performance domain in focus. Behaviors such as expressing confidence in members' actions, where the leader explicitly reinforces members' activities, can provide members both with insight into likely-to-be effective approaches, and also with momentum and self-assurance in taking future actions beyond any specific guidance provided by the leader.

In broad strokes, empowering leadership encompasses a set of behaviors that leaders enact that can have a systematic impact on team members' psychological empowerment. This can, in turn, impact members' feelings of ownership and enthusiasm for the team's collective aims. An important implication of empowering leadership is that, as the team struggles to identify suitable managerial and analytical Objectives, for example, members of the team build and retain feelings of ownership for efforts invested in each stage of the OUtCoMES Cycle. They will be more likely to invest in each stage of the discovery process, rather than passively allowing it to unfold under the authority of distal leadership. This is important as it increases the likelihood that a broader cross-section of members' experiences and inputs will be devoted to this critical process. Members will be more likely to dig deeply into the mechanical options available to advance project Objectives when they feel responsibility for the success of this process, and autonomy in establishing the trajectory of the Cycle as it advances.

7.1.3 Collective Leadership

While empowering leadership reflects an emphasis on the kinds of leader behaviors that provide project members with feelings of autonomy and responsibility for project work, collective leadership encompasses the architecture of members' interactions with one another. Specifically, collective leadership is a dynamic, interactive influence process reflecting the flow of various resources from member to member, including ideas, creativity, feedback, and support, among others.

Through these interactions, members exert social influence on one another, with an ultimate focus on the achievement of the team's collective aims. Under collective leadership, critical decision making and control are fundamentally distributed throughout the team as a consequence of the pattern and content of members' interactions with one another. This process reflects an on-going exchange of social and professional resources, and influence between the members of the team. The gravitational center of collective leadership, rather than being situated within a particular leader or cadre of individuals afforded authority by a leader, is actually present in the architecture of the interactions between members themselves.

Central to the concept of collective leadership, multiple team members simultaneously engage in the enactment of leadership through their interactions with one another. The structure of these interactions can be defined in a number of ways. These include how distributed these interactions are across all of the members of the team, and the density of these interactions across all possible interactions between members. When members' interactions are evenly dispersed across the team, this reflects what has been referred to as 'decentralized collective leadership'. When collective leadership is decentralized, the broadest possible cross-section of members' perspectives and insights are positioned to play a role in the approaches and perspectives adopted by the team, bearing directly on, for example, emergent definitions of Objectives, Utilities, and Connections.

The density of collective leadership, in contrast, reflects the level of involvement of members of the team. Often it is operationalized as a relative enumeration of interactions, scaled by the totality of possible interactions within the team. When collective leadership is dense, members have many interactions with the broadest cross section of other members within the team. Under such circumstances, the exchanges between members are at their richest and most informed, potentially yielding the most in-depth and potent insights into the nature and attributes of the team's set of candidate Objectives, Utilities, and Connections. This also fosters more intense and varied approaches adopted by the team, and thus a greater likelihood of successful innovation.

Another core attribute of collective leadership is that members of the team are likely to play multiple and distinct roles within the network of exchanges that define the architecture of the team's collective leadership. These roles have been serially defined as the Navigator role, the Engineer role, the Social Integrator role, and the Liaison role. For each of these roles, distinct activities and behaviors are enacted, distinct social resources are shared, and discrete aspects of the team's collective work are facilitated.

For example, execution of the Navigator role facilitates the team's development and maintenance of its principal Objectives, and the trajectory of its activities over time. Execution of the Engineer role, in contrast, helps provide structure to the team's project work (think Connections and Manifest), and facilitates coordination of members' collective efforts, helping to integrate disparate domains of expertise in the service of the team's collective tasks. Execution of the Social Integrator role helps to facilitate productive, supportive, strong social relationships between team members, that collectively allow the team to persevere through difficulties and turbulence, overcome intellectual and conceptual obstacles, and draw strength from one another in the face of crises, problems, and setbacks. Finally, the execution of the Liaison role facilitates the development and maintenance of functional and constructive working relationships with principal external interests and stakeholders. This allows the team to continually recalibrate and maintain focus on key project objectives and deliverables without losing sight of important external benchmarks core to defining team success.

7.2 Collective Responsibility

The fine points of distinction between these various depictions of group, project, or team-level leadership approaches serve to accentuate the importance of a broad underlying takeaway. Namely, effective leadership emerges through the distribution of the accoutrements of the leader's role to the individuals most in position to drive collective outcomes – the team's membership. This is essential, regardless of whether we are talking about (a) generating a shared sense of mutual responsibility for identifying the most appropriate Objectives and Utilities, (b) allocating distributed responsibility to members of the team for working through and consolidating Connections, or (c) deriving and evaluating, through interactions of team members, actionable models, intelligence and solutions.

In light of the complex dynamism that defines the work of engineering and analytics project teams, effective team leadership cannot simply flow downward from a project manager or leader to the members of the team. Success in this context fundamentally depends on the dispersion of responsibility to, and through, all of the members of the team. It is only as a mechanical consequence of the distribution of leadership in this way that members' disparate domains of knowledge, expertise, and experience are likely to be successfully leveraged through the stages of the OUtCoMES Cycle. Accordingly, it is against the backdrop of this kind of leadership sharedness within the team that we are able to fully consider the structural attributes of the knowledge and information architecture within the team as a whole. This composite architecture, its understanding, and potential for being effectively wielded, is an essential driver of a team's ability to access and leverage task-critical knowledge and information maintained by individual team members.

7.3 Memory and Learning

Repeatedly emphasized throughout this book has been the importance of successfully accessing, and leveraging, the wide-ranging knowledge and expertise housed within project team members, drawn from a cross-section of conceptually adjacent disciplines. Only when employed and harnessed collectively can the knowledge and expertise housed within the members of the team ultimately generate success. Of course, the complex and dynamic demands of a wide variety of engineering and analytics projects, coupled with the disparate, and disparately housed, expertise critical to various stages of the OUtCoMES cycle, create a classic challenge.

Members of the team, individually, don't possess the same knowledge and expertise as one another. They also are not likely to have all of the necessary knowledge and expertise to execute key project tasks and activities without the involvement and input of the other members of the team. The knowledge and expertise of all of the members of the team are critical to generating collective project success at various stages of the Cycle. This challenge creates a critical tension driven by the need for members without specific domain knowledge and expertise to both know the location of that knowledge and expertise within the team's membership and also to have timely access to it when faced with relevant project demands. The architecture of teams' collective knowledge, skills, and abilities is encompassed by what has been broadly depicted in the organizational sciences as a team's transactive memory system (TMS).

7.3.1 Transactive Memory Systems

A transactive memory system (TMS) is a collective knowledge infra-structure that enables teams to encode, store, and retrieve knowledge and information retained in the individual memories and experience repertoires of the members of the team. As noted in an earlier chapter, encoding is the process through which new project-relevant knowledge and information is drawn into the collective intellectual space in which team members' understanding of the project is embedded. This knowledge and informa-tion can be drawn from various sources, internal and external to the team.

External sources can include outlets such auxiliary organizational data bases, the internet, relevant trade publications, academic resources (e.g., published conference presentations; journal articles), knowledgeable friends and acquaintances, etc. Internal sources can include individual experience and information retained by members. Through these cached sources of information, individual members can add to the architecture of the team's mental models. Specifically, by way of encoding, this creates mental representations that can prove useful in both the identification of Objectives and Utilities, as well as in the structuring of Connections, the Manifest stage, and beyond. It becomes associated with existing knowledge housed within the team's collective memory, which in turn creates memory traces. Once memory is encoded, memory consolidation also facilitates storage in the team's collective memory, making it available for later use. This typically involves reorganizing memory traces and integrating them within the team's existing collective memory, making retrieval possible.

Importantly, in teams with a functioning transactive memory system, team members have a thorough understanding of which of the members of their team have the most developed, in-depth, and sophisticated knowledge and information about various aspects of the team's project work. They also possess an understanding of who is likely to be the most effective at completing tasks, duties, and responsibilities associated with various aspects of the project. This is referred to as meta-knowledge. Teams with a functioning transactive memory are thus positioned to assign various aspects of the team's tasks and responsibilities to the most qualified members when their knowledge and expertise are likely to be leveraged to the greatest effect.

What members also are positioned to do when TMS is operative is to seek out advice and insight from the members of their team who are the most knowledgeable about a given task at hand. Transactive memory systems can help project team performance because this knowledge infrastructure facilitates members' access to a significant volume of task-relevant knowledge and information, which is critical for project work. TMS grows and develops over time as members of project teams gain experience working with another, and through these experiences learn how best to .divide task-anchored responsibilities for learning, remembering, and communicating project-relevant information to other members of the team. As a result, TMS situates the most task/activity-anchored experts to play the most prominent roles in these stages of the OUtCoMES Cycle.

🌀 *Missing Links*

Felipe manages a small team of software engineers. When his company undergoes a restructuring, Felipe is suddenly made responsible for an additional software system that his team did not originally build, but that it does depend on for a key piece of functionality in the new software they have been actively developing.

His team is already overburdened and ill-equipped to handle the workload increase, but the restructuring was deemed necessary to reign in company costs. Most of the team is already assigned to a flagship goal of great concern to the company's CEO. Not wanting to delay the delivery of the team's first major milestone for this goal, Felipe assigns the task of facilitating knowledge transfer about the new software to a junior engineer on his team, Cody.

Cody does his best to learn as much as he can about the new software system in the brief period of time allotted for this work. While the rest of Felipe's team works toward the flagship goal, Cody is exclusively assigned responsibility for addressing all bug fixes and other tasks relating to this newly acquired software system. Over time, Cody is slowly able to build up his expertise with this new software. However, because all of the rest of the members of the team are fixated exclusively on advancing work toward the team's flagship goal, Cody does not share progress or insights into this work with the other members of the team.

During Felipe's product demo at the first milestone of the flagship goal, the CEO points out a critical failure observed in the new product experience. Frustrated, she asked Felipe to immediately perform a root cause analysis and come in with a revised, fully working demo by the end of the following week. After some time, Felipe's team traces the failure to a problem with the dependent software system they recently acquired which had been exclusively maintained by Cody. Unfortunately for the team, Cody is now on vacation, unavailable while visiting with his family outside of the country.

Felipe tells all of the other team members that this is an "all hands on deck" situation in order to fix the problem in the dependent software system. The team flounders and is unable to diagnose the issue in a reasonable timeframe because they just don't have any real knowledge about this new software. Ultimately, Felipe is forced to backtrack with the CEO, slipping the first milestone deadline, and rescheduling the demo three weeks into the future. No real progress can be made in diagnosing the product experience failure until Cody is back online, and can assist the rest of the team in isolating and resolving the underlying issue.

Learning Concept: **Communication Network** – The formal and informal linkages through which members of a team communicate with one another.

Reflection Questions

1   Why did knowledge of the team's software systems get siloed in this case? Was this the best approach for addressing this situation overall?

2   What mistakes did Felipe make in the communication network he structured for his team?

3   How might Cody have kept the other members of the team updated with details on the new software even while the rest of their teammates were focused on the flagship goal?

Research Follow-up – Examine the literature on alternative communication networks in teams. Try to identify when different types of structures should be used in different task situations.

7.3.2 TMS and Project-Critical Action

Transactive memory systems function broadly as a vehicle to facilitate the efficient integration and distribution of team members' knowledge and expertise. The system can be thought of as a cognitive warehouse, with a transparent and efficient inventory management platform, that allows team members access to the knowledge and informational resources they need to execute project tasks and activities effectively. However, beyond these operational mechanics, there are several specific ways that the presence of a functioning transactive memory system can help project teams to accelerate their arrival at an optimal, or nearly optimal, definition of the project's problem space and an identifiable path toward accomplishing key project Objectives.

Novel information. Among the most important success factors for analytics and engineering project teams is that the information flowing into the team reflects the most relevant and up-to-date intelligence bearing on all aspects of the project, across its phases of evolution (c.f., Thomke 2001). Stagnant, redundant, out-of-date, incomplete, or inaccurate information, at any of the phases of the team's development, can lead to a number of key process and outcome failures. For example, among the primary responsibilities project teams have at the earliest stages of the OUtCoMES Cycle is the identification of key project Objectives, and the development of a blueprint bearing on how these Objectives will be accomplished, reflected in Utilities. When TMS is operative in a project team, the filtering mechanisms that focus member-experts' attention on relevant external information are finely tuned.

Importantly, these filtering mechanisms are operative for all of the members of the team, who each maintain a slightly different understanding of what the project means, how its moving parts fit together and operate, and the relationships that define the various elements of the project. As a consequence, members of project teams with a well-developed TMS adopt slightly different frames of reference when thinking about aspects of the project that are relevant for the identification of Objectives, and how these might be accomplished. This orientation-specific variation in how members' information filters are tuned decreases likely redundancies in information in-flow as member experts will be searching external sources using different channel frequencies. TMS also increases the probability that information inflow will be more up-to-date (i.e., experts are closer to the source than non-experts), relevant (i.e., experts have a more well-defined understanding of specific aspects of the project than non-experts), and complete (i.e., experts will have a better sense of the range of relevant parameters in focus than non-experts). Thus, while TMS provides members with a transparent architecture that facilitates access to knowledge and information stored within the team itself, it also provides teams with access to novel information from outside the team.

A Culture of Innovation. Implicit to the OUtCoMES Cycle's emphasis on alternatives and critical examination (Scrutinize) is an openness to innovation that yields novel questions, new approaches, focused criticisms, and timely insights. An overarching culture of innovation, whether explicitly supported by the broader organizational context or not, can help teams to repeatedly, and authentically, peel back underlying assumptions, institutional inertia, and tacit (albeit extremely powerful) anchoring premises and theories that could otherwise inadvertently lead to front-line operator myopathy. Insufficient examination and analysis, coupled with premature consensus and acceptance can lead to expensive, time-consuming, and ultimately unfruitful tangents that can be difficult to extricate from. Linearity in conception and approach definitionally precludes the kind of exploration that plays a central role in the successful execution of the S-A3 through the stages of the Cycle.

In addition to all of its other virtues, TMS also has the ability to facilitate the development of cultures of innovation within teams. Because the responsibility for maintaining, and providing access to, distributed expertise within the team is housed within different member-experts, what emerges through activity and task-focused member interactions are explicit depictions that require not just domain specific logical anchorage, but also cross-domain logical consistency. Members are forced to dig deeply into their logical frames of reference when building out arguments or explanations, offering advice, or providing insight to other members who are drawing on their domain-specific knowledge and expertise.

The process of working through a position, idea, or suggestion (e.g., by way of S-A3 documentation) at the level of clarity necessary to generate value for non-domain experts, and experts with tangential/adjacent expertise, generates repeated opportunities for multiple perspectives and interests and frames of reference to be introduced and tested and challenged and discussed. Thus, the use of S-A3 documentation and the power of a functioning TMS, not only benefit teams as a consequence of a culture of innovation, but they also foster that culture's germination and development toward discovery.

7.4 Driving Transactive Memory in Project Teams

What is clear is that there are a number of ways in which an operative transactive memory system can help analytics and engineering project teams to generate functional long-term outcomes of the most direct and immediate benefit to stakeholders. TMS helps to provide project team members with the most useful knowledge and information because the filtering mechanisms that member-experts adopt as information flows into the team decreases information redundancy, and increases information freshness, relevance, and completeness. TMS also helps teams to leverage creative and innovative framing in building out definitions and approaches because the distribution of expertise across members introduces definitional transparencies and repeated opportunities to test, challenge, and discuss multiple perspectives and frames of reference. Finally, TMS also helps to provide project teams with a foundation for developing mature analytical models because it facilitates members' access to the most relevant, sophisticated, and up-to-date knowledge and expertise.

While there are several approaches that teams and team leaders can adopt to drive a functioning transactive memory system, the majority of these ultimately depend on the length of time that members of teams are together, and the experience of repeated exchanges that generate the specialization, credibility and coordination that are hallmarks of a functioning transactive memory system. However, these conventional, time-anchored approaches are not available to newly formed analytics and engineering project teams because they have not had the time or experience necessary to develop an understanding of the architecture of members' knowledge and expertise. Positioning project teams to benefit from a functioning TMS from the earliest stages of their inception ultimately depends on accelerating the pace at which knowledge of this architecture emerges across members.

If the members of the team don't have a coherent and consistent working trajectory that allows them to focus their efforts in the same direction, this can lead to members working at cross purposes, impeding the team's momentum and upside performance potential. Meeting complex project Objectives will always be a challenge. The pressure to meet Objectives only increases in intensity as decisions are made and resources are committed. As stakeholder expectations increase, timelines shorten, budgets are constrained, and competition increases, these factors only increase the pressure on teams to find ways to fully leverage the team's resources. In successful teams, the operational and process aims of the team align with members' knowledge, skills, abilities, and other characteristics (KSAOs).

While leaders can micromanage the process of generating an alignment between the needs of the team and the KSAOs of the teams' members, this can diminish members' feelings of autonomy and responsibility. Members are likely to have higher levels of motivation when they can take ownership of the team's Objectives, and importantly, when leaders can empower members to actively match the team's performance and process goals with their own KSAOs.

The members of project teams all bring different skill sets and experiences to the team, and it is a reality of project work that some members are better at some aspects of the team's activities than others. Knowing who can do what within the team, reflective of an operative transactive memory system, can help to systematically align members' KSAOs with the team's aims across the stages of the Cycle. Transparent, distributed knowledge of members' KSAOs provides a mechanical foundation for the development of a functioning transactive memory system.

Catalyzing the development of the team's cognitive warehouse depends on gathering and codifying accurate depictions both of members' understanding of the team's goals across the stages of the Cycle, as well as members' KSAOs that correspond with these goals. Anchored within an easy-to-use exercise (e.g., T-GAME – Team Goal Ability Matching Edge; see www.masteringdiscovery.com), the updated architecture of members' KSAOs and team Objectives, can be continuously shared with all of the members of the team, in real time, during the course of project work. The framework depicted in T-GAME, for example, reflects the full spectrum of Objective-anchored (and updatable) knowledge and informational resources available within the team. This resource represents a technological substitute for the meta-knowledge on which a functioning transactive memory system depends, and which in typical team contexts takes months (or more) of collaboration to develop and leverage in an effective way. As such, an easily accessible, virtual warehouse accommodating a comprehensive source-book of members' KSAOs, coupled with an explicit connection to the stages of the Cycle, provides an additional vehicle for generating competitive advantage among engineering and analytics projects teams. Central to the effective leveraging of this approach is the maintenance of a dynamic compendium of KSAOs, analogous to project details tracked and revised via S-A3s, across the stages of the Cycle. Such a dynamic, living virtual architecture serves as a powerful mechanism to expedite value from TMS.

Practitioner Recap

The complexity of analytical and engineering problems requires thoughtful and intentional development of a leadership style that provides team members the right to make decisions, with the responsibility that those decisions advance the work toward realizing the project's Objectives. Single-treaded, top-down leadership styles have proven to be ineffective in this context. Shared leadership, Empowering leadership, and/ or Collective leadership are effective in these complex project environments. Collective responsibility depends on dispersing leadership throughout the team to leverage the embedded knowledge and skills of all team members, required for different stages of The Cycle. A team's Transactive Memory System (TMS) helps members to encode, store, and retrieve knowledge and information retained by individual members. Time and experience can help to further develop TMS, but new teams need accelerated methods to accomplish the efficiencies available through TMS. Codifying and sharing members' knowledge, skills, abilities, and other characteristics (KSAOs) and team Objectives in a framework like the T-GAME has the potential to accelerate TMS development. The T-GAME outlines Objectives across The Cycle stages and the necessary member KSAOs to achieve them. Continuously updating this living record, like revising S-A3s, helps to build TMS to generate competitive advantage. A virtual warehouse of dynamic KSAOs and Objectives data can facilitate expedient TMS development and use in both new and experienced project teams.

Conclusion

Discovery Mastered

There's a tendency to refer to the front end of projects as "fuzzy". This depiction isn't unjustified. There are a host of critical issues that are unavoidably characterized by uncertainty as these begin to unfold. Indeed, the full set of issues necessary for consideration is itself ambiguous at the outset of a project. There's a lack of clarity regarding the ways in which resources could most profitably be deployed, and how a team's time should be spent. Suggesting that this fuzziness exists isn't an excuse. It isn't a defensive game that project managers play in order to hedge against the risk of bad results. It's a reality.

However, while the ubiquitous presence of a wide range of unknowns doesn't necessarily condemn the vast majority of projects ... it is also something not best left to guesswork or blind faith.

In this book, we have repeatedly referenced two key concepts, *structure* and *alternatives*, both of which are critical to discovering ideal project opportunities and seeing them through. Through much of our general treatment of the broad topic of problem identification, we have emphasized structure explicitly: structured processes such as The OUtCoMES Cycle, structured documentation such as the S-A3 and structures for managing team development and productivity. However, in our discussion of each of these elements of structure, we have also explicitly emphasized the value of retaining and returning to alternative options as a way to gauge performance, and also to account for critical relationships in design. There is obviously non-trivial value in defining the space we work in. But there is likely equivalent value in remaining unbiased to in-real-time adjustments of that space. In doing so, in allowing the proverbial doors to remain open, we can maximize our chances of learning things that can help; not only things of relevance to the project at hand, but also of relevance to future work down the road.

The world is a very noisy place. The more we can do to transform that noise into the kind of variation we *can* make sense of, explain, and account for, ultimately the more equipped we are likely to be to deal with it

DOI: 10.4324/9781003427650-11

proactively and effectively. An identical set of mechanics are at play for the front end of projects. It is the inherent fuzziness that defines a project's beginnings that we target most directly with our approach. It is the underlying "unfuzzying" that we describe here that has yielded the substantive boosts in clock speed and project value that clients we have worked with have appreciated the most. The increased clarity, specificity, and focus that emerges from this discipline is why clients continue to leverage The OUtCoMES Cycle in subsequent projects. It is why they continue to invest time to develop novel and meaningful performance metrics that bridge managerial and analytics interests. It is why they have made documentation such as the S-A3, and project team leadership concepts such as shared leadership, collective responsibility, and transactive memory systems (TMS) components of their common tool set and vocabulary.

Once invested stakeholders have made use of The OUtCoMES Cycle, do they immediately discard all of the valuable knowledge and tactics accumulated through hard-earned experience and learning with PDCA, DMAIC, Double Diamond, and Design Thinking principles in general? Of course not. And, as invested stakeholders ourselves, we would never suggest that they should do so. These experiences are an integral part of the collective institutional memory and tacit knowledge held by managers, engineers, analytics, and project leaders. Embedded in these anchored instances and process experiences are extremely useful reference points and rules of thumb. The OUtCoMES Cycle, as we've sought to demonstrate in a clear and linear way throughout this book, captures the key elements of these approaches as well. The Cycle capitalizes on the best elements of these approaches, while concurrently drawing attention to – and addressing – critical questions that often are not sufficiently dealt with early on. The Cycle emphasizes both forward progress as well as retrospection. It highlights the importance of brainstorming, scrutinizing, and selecting Objectives, as well as the means of their pursuit; which are likely to provide an explicit intersection between problems, solutions, and users. Fundamentally, The Cycle also matches these considerations with a process that explicitly outlines how analytical success supporting any related solutions might be measured. In this sense, The Cycle continuously bridges the operational imperatives and dialects of management, engineering, analytics, and end users.

Can all of these ambitions be accomplished in the absence of The OUtCoMES Cycle. Yes, in the sense that riders in the Pelatonia or Tour de France might also be able to compete in these races with flat tires, poorly positioned seats, and blinders. However, the best teams would never handicap their riders in this way. Afterall, professional teams don't just want to finish the race. They want to win it! Companies are increasingly competing based not just on their ability to push out novel and impactful solutions, but to consistently do so significantly ahead of the competition. The bar keeps getting raised higher and higher. We can't continue to use the same old bikes that we rode twenty years ago. If we want to be adaptive, we need to adapt the processes on which our flexibility relies. This doesn't mean throwing out all of the tools that still work. But, it does mean thinking about these tools a little differently. Making smarter use of these tools, and adopting new tools, to replace those that no longer provide a competitive advantage.

As you move forward with your next project, we encourage you to take a moment of reflection to think about the performance space that defines the project. List and scrutinize the possibilities ahead of you moving into the earliest phases of the project. What are the meaningful measures of practical success for the project? Are there both upper and lower limits to that success? How close are you to those limits now? What quantifiable evidence will be available to suggest that you have a path toward advancing these outcomes? What are the means by which you might pursue these advancements? Are these levers subject to limits? Are the likely relationships connecting these levers and outcomes complex? Are they sufficiently understood? How can you quantify that understanding as additional evidence for moving forward? In this moment of reflection, in short, go through the questions that project leaders ask as they develop, document, and refine their Fundamental Managerial Objectives, Analytical Objectives, Utilities, and Connections.

When you get to the stages in your project where conceptual models of cause and effect begin to Manifest holistically, capture those models adequately for future reference. Detail additional evidence that emerges as you Explicate the evidence-based structure of these models; i.e., as you analyze and estimate these models. Scrutinize the relevance of these emerging findings with regard to all aspects of practical fit. Listen to those voices that suggested alternatives early on. Listen to your own internal skeptic.

And, critically, retain and share the wealth of your evolving experience through clear and collaborative discussions with all relevant stakeholders, and also through explicit documentation for future reference. Some of the best ideas for new projects come from those that were passed on due to timing, resource limitations, conflicting priorities, and other factors that kept them in a subordinate position. As Faulkner famously wrote: "... the past is never dead. It's not even past" (1951). Keep searching for the next discovery on the horizon, but never forget to look behind you before starting your next race.

Glossary of Key Terms

5W+H   The acronym reflects the five questions common to design thinking processes. The 5 Ws and the 1 H refer to, respectively, why, who, where, when, why, and how. The intention of these questions is to facilitate understanding of a given performance context and deepen insight into a problem or set of issues by digging into its key attributes. A method ostensibly useful in identifying FMOs, Utilities, and Connections within the OUtCoMES Cycle.

A3   See S-A3.

Bottleneck   A bottleneck, short- or long-term, is a point of congestion in a process or value-adding system that temporarily arrests or significantly slows the pace at which the system can operate. System inefficiencies that emerge as a consequence of the bottleneck can create delays, generate higher costs, and limit throughput and associated performance.

Bounds   Bounds or the boundedness of a performance context are the physical and logical limits that define a performance space.

Choice Criteria   Critical attribute or set of attributes used by invested stakeholders in evaluating a set of alternatives.

Clustering   An exploratory statistical analysis used to identify homogenous groups of cases, or structures, within a set of data.

Collective Knowledge   A state where a large group of affiliated individuals combine their knowledge and information in the service of a group or collective task, enhancing the capacity of the group to operate and function as a consequence.

Complexity   In project contexts, complexity is often described by objective features such as the number of elements involved, the independence of these elements, the uncertainty present in the project's Objectives, and the uncertainty present in the impact associated with project Utilities as they Connect to such Objectives.

Connections   In the OUtCoMES Cycle framework, Connections refer to relationships between and among Objectives and Utilities that capture cause-and-effect or coincident association with other factors, such as limits (constraints). Connections can have deterministic and mechanistic characteristics (anticipated by definition), as well as stochastic and seemingly random characteristics that rely on prediction and understood distributions in risk. Connections may involve seemingly simultaneous or lagged relationships, as well as feedback mechanisms involving subsequent reinforcement phenomena, or counter-balancing phenomena among Objectives and Utilities. See Chapter 3.

Constraints   Constraints encompass limitations on the capacity of a system to produce or distribute goods and services, and may include factors related to capacity, resources, bottlenecks, and demand.

Continuous Measure   While discrete measures typically capture unique and separable states of finite variety (e.g., yes/no, group1/group2/group3 determinations), a continuous measure theoretically encompasses an unlimited set of options, unbounded or bounded by (contained within) upper or lower limits (e.g., width, speed, temperature, height, weight, time).

Control   Both the means of evaluating and also the capacity to generate modifications and changes to key facets of an activity or process in order to effectuate positive changes in cost, scheduling, quality, or other factors having a negative impact on key performance outcomes.

Degrees of Freedom   Reflects that latitude present in a system or series of related systems that are amenable to adjustment in the service achieving measurable performance outcomes, within the limits of system constraints.

Descriptive   Descriptive analysis helps clarify "what" we are dealing with in a problem. We might think of this as 'describing our local reference system'. That is, the most effective descriptive analysis informs us regarding the most central issues that we are contending with, and those factors that may influence or limit those issues. See Chapter 1.

Design Thinking   Broadly, a common approach toward problem solving encompassing collaboration, innovation, and acceleration. Such thinking is implicitly embedded in the process and framework of the OUtCoMES Cycle.

DMAIC (Define, Measure, Analyze, Improve, Control)   A five-phase process improvement approach, with each stage mapping to critical aspects of the OUtCoMES Cycle.

Dominant Paradigm   The accumulated norms, beliefs, values, habits, patterns of interaction, and assumptions that shape the world view most commonly maintained within a given cultural context.

Doubling Down To continue investing time, energy, effort, and resources in the furtherance of a particular goal in a more determined way than before.

Equifinality A condition in which similar results can be achieved through the use of different approaches across different situations.

Escalation of Commitment Psychological tendency for decision makers to continue to invest resources, and even increase levels of investment, in a failing project rather than terminate it.

Explicate In the OUtCoMES Cycle framework, this action involves the derivation of actionable intelligence, based on a set of understood and interconnected rules and relationships. In predictive exercises, this involves the estimation of effects and predictive structures (i.e., predictive Utilities), whereas prescriptive exercises involve the assignment of value to decisions (i.e., prescriptive Utilities). Heuristic tactics, mathematical approaches, and computational algorithms are all fair game, depending on alignment with the specific modeling exercise in question. See Chapter 4.

FMOs (aka Fundamental Managerial Objectives) Objectives that are both Fundamental in their prioritization and Managerial in that they have direct bearing on key performance measures and relevance to stakeholders beyond the project team. See {Objectives} and Chapter 2.

Failure Across multiple sets of stakeholder expectations, project failure is declared when promised, expected, or anticipated goals, outcomes, timelines or budgets have not been met.

Feasibility Determination based on a comprehensive evaluation of all critical factors central to a project to determine its likelihood of being successful.

Firefighting Leadership style where the emphasis is less on proactively addressing anticipated situations and more on addressing immediate problems as they arise.

Fishbone/Ishikawa Diagram See Relative-Impact Fishbone.

Fit The third major criterion for FMO selection, relates to whether pursuit of any given candidate Objective is in fact aligned with all of the other moving parts in the system, including commitments of time and resources that might otherwise be allocated differently.

Functional Boundaries The points at which the knowledge, skills, abilities, and other attributes associated with work in one functional area of an organization intersect with the set of KSAOs in another area of an organization.

Fuzzy Front-End Predictive, pre-development activity that occurs during the period between initial project inception and project initiation.

This activity, at its best, is based on informed guesswork and systematic study; Encompassing the identification of problems with sufficient transparency, plasticity, and fit.

GMM   A three-part statistical approach combining observed data with the ratio of the size of a population at an emergent equilibrium with the size of the initial population (e.g., a population moment), where a model is depicted in the form of a population moment, model parameters are estimated using the population moment, and then generating statistical tests of the validity of the model and model parameter efficiency.

Hamiltonian Cycles   A Hamiltonian cycle, or a 'tour' is a path through the points on a graph that begins and ends at the same point and that also includes every other point on the graph exactly one time. See Icosian Game.

Heuristics   A broad decision-making approach or 'rule of thumb' that is often used to quickly generate effective solutions to the allocation of resources or selection of best options, albeit often without the guarantee of optimality in such decisions.

Icosian Game   A mathematical game, invented in 1857 by William Rowan Hamilton encompassing the task of finding a tour, or path including every vertex in a Hamiltonian graph.

Impact   Impact reflects how a project affects a wide range of stakeholder interests including the environment, the organization, and the community. It encompasses positive and negative, primary and secondary, long term and short term effects.

Interval Measure   A quantitative variable used in statistical analysis, defined by a consistent unit of measurement, that represents a range of values, where the difference between any given values along that range is meaningful.

Iterative Feedback Process   A technique intended to improve an idea or process design through repeated consideration by another party or parties, and subsequent adjustments by the idea or process design originator(s).

KSAO   Acronym standing for Knowledge, Skills, Abilities, and Other characteristics that reflect the attributes of key project personnel of relevance for execution of project-related tasks, duties, and responsibilities.

Limitations   In the context of project management, any given restriction imposed, either internally or externally, on the options available in the execution of a project. See Constraint.

Living Document   Often referred to as an Evergreen document or a Dynamic document, any project-related codification that is continually updated, edited, or modified as the project progresses through its stages.

Mental Model   Reflects an underlying or inherent frame of reference or experiential lens that is used repeatedly in the development of strategies or approaches for solving problems. See Reference System.

Manifest   In the OUtCoMES Cycle framework, this action involves the assembly of the full set of Connections between Utilities and Objectives, towards the formation of an integrated system of relationships. The product is a system of interconnected rules and relationships that essentially define the scope of the decision space for analysis and design. It creates the foundation on which depend the exploration and search for solutions that fit both the problem specifications and the context for application. See Chapter 4.

Monotonic   A function or model that is either always entirely always increasing, or always decreasing.

Objectives (and Candidate Objectives)   In the OUtCoMES Cycle, Objectives describe measurable outcomes that serve as both the motivating force catalyzing project investments, and around which local reference system details orbit, as well as definitive yard sticks by which project success is ultimately measured a posteriori. Objectives are subdivided into Managerial and Analytical forms, crossed with either Fundamental or Means orientations. Managerial Objectives are those that have a direct connection to managerial practice (e.g., anticipate errors more easily, or reduce error rates), whereas Analytical Objectives relate to the objective goals of analysis in support of these (e.g., maximize log-likelihood or accuracy of a predictive model, or minimize errors subject to resource constraints). Fundamental Objectives, Managerial or Analytical in nature, are the highest level in a project and represent the overall motivation of a project. Means (or Intermediate) Objectives represent critical Managerial and Analytical stepping stones that must be achieved while pursuing Fundamental ones (e.g., descriptive work precedes predictive, which typically precedes prescriptive). Objectives considered during early brainstorming and in later project stages (where project aims may be reassessed) as potential areas of focus (e.g., Fundamentals to be) are referred to as Candidates. See Chapter 2.

Omissions   Mistakes and errors in design involving the absence of critical data, requirements, perspectives, or analysis, particularly when they relate directly to Objectives, Utilities, or Connections.

Operational Configuration   The final set of attributes and characteristics, including all functional and physical specifications, that ultimately define a project's product or deliverable (i.e., the final system for which Objective performance is gauged).

PDCA (aka Plan, Do, Check, Act)   A classical framework that project managers have used to systematically implement incremental change- and motivate continuous improvements in performance.

Plasticity   A characteristic of the most impactful of FMOs, wherein variance in Objective performance has been documented or otherwise validated, and there is some foundation for developing a plan of action to deliberately achieve desired changes.

Predictive   Predictive analysis emphasizes explication of the role played by Utilities and the form/nature of their associated Connections to Objectives. Prediction can be accomplished using an array of tactics, from basic linear regression to more sophisticated machine learning approaches for classification. These tactics facilitate understanding of not only why things may have happened in the past, but also provide guidance for movement toward achievement of future objectives. See Chapter 1.

Prescriptive   Prescriptive analysis explicitly emphasizes the operational "how", where the intelligence generated from descriptive and predictive analysis is used to formulate a systematic solution. While descriptive and predictive analysis are critical to piecing together the puzzle (i.e., structuring the problem), prescriptive analytics facilitates the development of tactics, options, and limitations necessary to navigate the twists and turns of the maze that emerges from this process (i.e., development of practical solutions). See Chapter 1.

Priorities   The determination of, based on a set of stakeholder, organizational, and project capability criteria, an ordinal set of potential options sequenced with those most critical and viable ahead of others.

Problem Statement   Explicit, brief, depiction of an unclosed gap, an unachieved outcomes or an yet to be surmounted obstacle encompassed by the domain of a project. In short, such a statements should articulate the Fundamental Managerial Objective, among other subordinate Objective details.

Programmed Decisions   Routine, repetitive decisions that follow an established pattern and set of guidelines, and that have a set of rules in place that render decisions routine.

Project   A project is an entity defined by two key characteristics: (1) its intent is to be temporary, with limited scope (in contrast to an ostensibly sustainable operational processes), and (2) it is undertaken in to generate a targeted deliverable such as a product, service, process change, or other outcome.

Psychological Safety   The subjective belief that one is safe from punishment or humiliation as a consequence of speaking up with ideas, questions, or concerns, in a group or team reflects this shared expectation across teammates.

Reference System   A clear, unambiguous, and confined specification of elements, rules, conventions, relationships, and numerical constants used to help focus project discussions and priorities. See Scope.

Relative-Impact (RI) Fishbone   A visual brainstorming approach used to identify a range of possible drivers of a given problem, and to sort these drivers into problem-relevant cause-and-effect categories.

Risk   Uncertainty surrounding a situation, event, or condition that, if it materializes, can have a positive or negative impact on the outcomes associated with a project.

Robustness   Reflects a circumstance such that the Objectives of a project will be reached or accomplished despite the emergence of undesirable or unexpected deviations from established plans.

S-A3   A documentation structure designed to depict the nature of an issue or a problem and various approaches toward addressing them. The documentation follows the stages of the OUtCoMES Cycle. Confinement of documentation to a single legal paper area, mandates priorities in such documentation, such that non-critical information is constantly subject to replacement. The retention of earlier versions of S-A3s ensures a documentation history for later reconsideration and forensics.

Scapegoated   Opportunistically singling out a member of a project team following a failure or setback of some kind, typically a member who has contributed visible input at some point in the process of executing various aspects of the project, as the underlying cause or driver of the failure.

Scope   A foundational aspect of project implementation that facilitates determination of project goals, constraints, workflow management strategies, project tasks, and deliverables.

Scrutinize   In the OUtCoMES Cycle, this action involves closely and pragmatically examining outputs of the prior stages, prompting corrective redress and reconsideration of alternatives when fit among the problem, solution, and user is neither obvious nor sufficient. Although positioned at the end of The Cycle, more realistically, consistent with their respective descriptions and the tactics recommended for executing prior stages, scrutiny pervades each stage. See Chapter 4.

Specification   Documentation used for project management that holistically defines a project's management plan, accurately specifying project needs (including Objectives), constraints, expected features, deadlines, and budgeting constraints.

Stage   Major within project executive-level control points, the purpose of which is to keep the project on track.

Stakeholder   Individuals and entities directly or indirectly impacted by the project, and in position to impact the trajectory of the project, at any point in the project's lifecycle.

Status Quo   Accepted processes or procedures.

Sustainability   A project management approach that seeks to systematically balance environmental, social, and economic interests in order to meet the needs of both current and future stakeholders.

Systems Thinking   A holistic analytical approach focused on capturing and appreciating the dynamic manner in which the constituent elements of a system interrelate to one another over time, and their operation within the context of the larger systems in which they are embedded.

Tactic   Instantiation of an action plan, reflected in the specific actions taken to execute the plan.

Target   A documented set of established Objectives that define how any given project should be done, and the results or outcomes that the project is expected to generate.

Tradeoffs   Challenges and choices in resource allocations, where it is presumed that one or more options must be sacrificed for the advancement of others. Tradeoff decisions in project management aim to maintain a functional balance between a project's schedule, budget, and performance, while relinquishing at least one preferred option or alternative.

Transactive Memory   A knowledge infrastructure that allows teams and groups to encode, store, and retrieve information.

Transparency   Transparency is a condition where all project-relevant information is shared with project stakeholders including project goals, scope, timeline, and budget. It is also a trait of the most effective Objectives (specifically the most effective FMOs), in that it describes the ability to measure both current and prior states, thus opening the door to both analysis and benchmarked improvement.

Utilities   In the OUtCoMES Cycle framework, Utilities describe options that decision-makers have control over (either direct or indirect control), and which through that control have the potential to advance specific Objectives. They can take the form of decision variables to be optimized, or coefficients in predictive models to be estimated. Like Objectives, the conceptual identification of Utilities draws on real-world considerations, while their numerical specification benefits greatly from analysis. See Chapter 3.

Visualization(s)   A process using, and artifacts based on, the graphical rendering of critical related data elements. The aim is to provide simple and clear representations of otherwise abstract patterns of relationships, to help provide insight from data, and communicate these insights both within project teams as well as to external stakeholders. See Bendoly and Clark 2016.

References

Basole, R., Bendoly, E., Chandrasekaran, A., & Linderman, K. (2022). Visualization in operations management research. INFORMS Journal on Data Science, *1*(2), 172–187.

Bendoly, E. (2016). Fit, bias and enacted sensemaking in data visualization: Frameworks for continuous development in operations and supply chain management analytics. Journal of Business Logistics, *37*(1), 6–17.

Bendoly, E. (2019). A Framework for Analytical Approaches. Whitepaper, published by the International Institute for Analytics.

Bendoly, E. (2020). Excel Basics to Blackbelt: An Accelerated Guide to Decision Support Designs, 3rd Edition. London: Cambridge University Press. (ISBN: 978-1108738361).

Bendoly, E., & Clark, S. (2016). Visual Analytics for Management: Translational Science and Applications in Practice. London: Routledge. 2016 (ISBN: 978-1138190726).

Bond, S. D., Carlson, K. A., & Keeney, R. L. (2008). Generating objectives: Can decision makers articulate what they want? Management Science, *54*(1), 56–70.

Cameron, J., & Bryan, M. (1992). The Artist's Way. Los Angeles: Jeremy P. Tarcher/Perigee.

Cox, J., & Goldratt, E. M. (1986). The Goal: A Process of Ongoing Improvement. Croton-on-Hudson, New York: North River Press.

DeTreville, S., & Browning, T. (2023). A fast-and-frugal heuristic lens view of the Toyota production system. Journal of Operations Management.

Drucker, P. F. (1954). The Practice of Management. New York: Harper.

Edmondson, A. C. (2004). Learning from mistakes is easier said than done: Group and organizational influences on the detection and correction of human error. Journal of Applied Behavioral Science, *40*(1), 66–90.

Faulkner, W. (1951). Requiem for a Nun. New York: Penguin Random House.

Fuller, B. (1975). Synergetics: Explorations in the Geometry of Thinking. New York: Macmillan Pub Co..

Gigerenzer, G., Todd, P., & ABC Research Group. (1999). Simple Heuristics that Make Us Smart. Oxford: Oxford University Press.

Goldman, B., & Taylor, P. (2023). Team Analytics: The Future of High-Performance Teams and Project Success. NY: Routledge.

Holmstrom, B. (1989). Agency costs and innovation. Journal of Economic Behavior and Organization, 12, 305–327.

Homer and Fitzgerald, R. (1961). The Odyssey. Translated by Robert Fitzgerald. Garden City, NY: Anchor Press/Doubleday.

Ionesco, E. (1969). Découvertes. Volume 3 of Sentiers de la création. Geneva: Albert Skira.

Kendrick, A. (2016). Scrappy Little Nobody. New York: Touchstone Books/Simon & Schuster.

Kim, D. H. (1999). Introduction to Systems Thinking. Innovations in Management Series. Waltham, MA: Pegasus Communications.

Lévi-Strauss, C. (1964). Le Cru et le cuit (The Raw and the Cooked). In Mythologiques. Translator John and Doreen Weightman. Paris: Plon Pub.

Lewrick, M., Link, P., & Leifer, L. (2020). The Design Thinking Toolbox. Hoboken, NJ: John Wiley & Sons.

MacDuffie, J. P. (1997). The road to "root cause": Shop-floor problem-solving at three auto assembly Plants. Management Science, 43(4), 479–502.

Meadows, D. H. (2008). Thinking in Systems. The Sustainability Institute. White River Junction, VT: Chelsea Green Publishing.

Muir, J. (1911). My First Summer in the Sierra. Boston: Houghton Mifflin.

Nallusamy, S. (2016). Productivity enhancement in a small scale manufacturing unit through proposed line balancing and cellular layout. International Journal of Performability Engineering, 12(6), 523–534.

Payne, J. W., Bettman, J. R., & Schkade, D. A. (1999). Measuring constructed preference: Towards a building code. Journal of Risk and Uncertainty, 19(3), 243–270.

Project Management Institute. (2017). A Guide to the Project Management Body of Knowledge (PMBOK Guide) (6th ed.). Project Management Institute.

Pyzdek, T. (2003). The Six Sigma Project Planner: A Step-by-Step Guide to Leading a Six Sigma Project Through DMAIC. New York: McGraw-Hill.

Saint-Exupéry, A. de. (1948). Citadelle. Gallimard.

Samuelson, P. A., & Crowley, K. (1986). The Collected Scientific Papers of Paul A. Samuelson, Volume 5. Boston: The MIT Press.

Sobek, D., & Smalley, A. (2008). Understanding A3 Thinking: A Critical Component of Toyota's PDCA Management System. Boca Raton, FL: Productivity Press/Taylor & Francis.

Thomke, S. (2001). Enlightened experimentation. Harvard Business Review, 79(2), 67–75.

Thompson, J. K. (2020). Building Analytics Teams: Harnessing Analytics and Artificial Intelligence for Business Improvement. Birmingham, UK: Packt Publishing Ltd.

Watanabe, K. (2009). Problem Solving 101. London: Penguin Books.

Appendix A

A Packing Exercise (Chapter 1)

Figure A.1 A packing solution that accommodates all 14 items.

Appendix B: About the Authors

Dr. Elliot Bendoly is the Distinguished Professor in the Operations and Business Analytics department of the Fisher College of Business, at The Ohio State University. Profession Bendoly has served as dean of the College's top-rated undergraduate business program, and has led the Specialized Masters in Business Analytics program. Before joining Fisher, Dr. Bendoly was the Caldwell Research Fellow and Associate Professor in

Information Systems and Operations Management at the Goizueta Business School of Emory University. His pre-academy industry experience includes work as a research engineer for the Intel Corporation. He holds a Ph.D. from Indiana University in the fields of Operations Management and Decision Sciences, with an Information Systems specialization in ERP and Knowledge Management. More recently he has been involved with coursework on modern analytics and visualization, IT-supported service operations, and DSS development for managers. The professor serves as Senior Editor at the *Production and Operations Management* journal (Behavioral Operations and Management of Technology departments) and as Department Editor for *the Journal of Operations Management* (Technology Management). His own publications in *POM* and *JOM*, combined with his works in *Management Sciences, Information Systems Research, MIS Quarterly* and *Journal of Applied Psychology*, and the *Journal of Business Logistics* represent no less than 34 published academic articles.

He has authored no fewer than an additional 33 articles in outlets including the *Decision Sciences, Journal of Supply Chain Management, EJOR, IJOPM,* and *Decision Support Systems*. His current research interests are split between studies into the effectiveness of Operations/IT alignment, and investigations in the Behavioral Operations domain: collaboration/group dynamics; and work policies/task complexity/uncertainty. Dr. Bendoly is the author of *"Excel Basics to Black Belt"* (Cambridge Press 2020; 2013; 2008), a book on which his Crystal Apple Award winning elective is based (www.excel-blackbelt.com). He is also the co-editor of *"Strategic ERP Extension and Use"* (Stanford Press 2005) and the *"Handbook of Research in Enterprise Systems"* (Sage 2010), the *"Handbook of Behavioral Operations Management"* (Oxford 2015), and *"Visual Analytics for Management"* (Taylor-Francis/Routledge 2017). Coursework: www.excel-blackbelts.com; Wikipedia: https://en.wikipedia.org/wiki/Elliot_Bendoly

Dr. Daniel (Dan) Bachrach is an Elected Fellow of the American Psychological Associate (APA), Society for Industrial and Organizational Psychology (SIOP), and the Association for Psychological Science (APS). He has won multiple awards for both teaching and research, including the University of Alabama's highest award for teaching, the National Alumni Association's Outstanding Commitment to Teaching Award, and the Culverhouse College Board of Visitors Research Achievement Award. Dan is a Full Professor of Management, and a Morrow Faculty Excellence Fellow. He is the author, co-author, or co-editor of 14 books on technology, sales, behavioral operations, and management, including two leading management textbooks, *Management* (15e) and *Exploring Management* (7e), used throughout the United States, with versions translated into Chinese, French, Greek, Indonesian, Portuguese, and Spanish used around the world. He has published more than 60 research articles, and his work regularly appears in the field's most elite scholarly journals, including *Organization Science, The Strategic Management Journal, the Journal of Applied Psychology, Organizational Behavior and Human Decision Processes, the Journal of Management, and the Journal of Operations Management*, among others. According to Google Scholars, Dan's work has been cited more than twenty thousand times to date.

Dan is an Associate Editor at the Journal of Organizational Behavior (Impact Factor 5.97), and a recognized expert in several academic areas, including Organizational Citizenship Behavior and Transactive Memory Systems in teams and leadership. Dan has received more than one point three million dollars in extramural funding from public agencies since 2017. His academic research has been iteratively funded by independent divisions of the United States Army, including the Research Facilitation Lab and the Basic Research Division of the Army Research Institute and the National Science Foundation. Dan teaches the mass Introduction to Management sections of Organizational Theory and Behavior to face-to-face classes of 250-plus students and mass, online sections in the Culverhouse College of Business. He also teaches the first-year Leadership and Ethics course in the MBA core, as well as courses in team productivity in the Executive MBA program and Executive Management programs at the Manderson Graduate School of Business.

Kathy Koontz is Analytics Platform Strategist with AWS, working to help customers create sustainable competitive advantage from their data and analytics investments. She has more than 30 years of experience working with large organizations in data and analytics across multiple industries including retail, financial services, public sector, and others. Her roles have been in both business and technology ranging from ETL developer writing SQL code, working as a marketing analyst measuring omni-channel campaign effectiveness, to leading analytic transformation programs at multiple Fortune 100 financial services companies. Additionally, she has demonstrated success in building high-performing teams, driving organizational change, improving analytics maturity, and leading customer-centric transformations. She is a regular conference speaker focusing on realizing sustainable competitive advantage from analytics and serves as an advisory board member for the Customer Analytics Program at The Wharton School, University of Pennsylvania. When not transforming analytics organizations, Koontz enjoys spending time with her husband and two daughters in a variety of activities including road-cycling yoga, snow skiing, hiking, cooking, traveling, and other adventures. She also raises money for cancer research through the Pelotonia community.

Porter Schermerhorn is a Principal Engineer, at Amazon. Porter initially joined Amazon Web Services (AWS), working to expose Amazon's underlying e-commerce platform to third-party developers. He led the development of AWS ROSE (Retail Ordering Services, U.S. Patent 8,627,379), a private web service for trusted 3 P partners which also served as the control plane for Checkout-By-Amazon (CBA). Porter next joined the Affiliate Marketing team where he helped design the Dynamic Throttling Web Service (U.S. Patent 8,281,382) and drove the re-launch of the Product Advertising API (PA-API) in the Amazon Affiliate Marketing Developer Toolkit. During his tenure in Amazon Consumer Marketing, Porter was the driving force behind many innovations, including CAFÉ (Common Automated Framework for Engagement, U.S. Patents 9,805,177 & 9,959,551). Porter then rotated from Consumer Marketing to serve as a floating technical resource supporting the Amazon Fashion division, focusing on Amazon's Merch on Demand business and its fashion subsidiary Shopbop.

Most recently, Porter has taken up a new position in Amazon's Global Media & Entertainment division, where he is chartered with defining the technical vision and strategy for entertainment-centric merchandizing and shopping experiences across media properties such as MGM and Amazon Studios. Early in Porter's Amazon experience, he found a pathway to add additional value in career mentoring and coaching, especially for early-mid career engineers. He is a longstanding member of Amazon's Interview Bar Raiser program and leads the Career Development content track at Amazon's annual internal Developer Conference. Along the way, Porter recognized the equal importance of the application of EQ alongside IQ in delivering durable success for both Amazon's customers and one's own career. Long a self-proclaimed student of luminaries like Daniel Goleman (Emotional Intelligence 1995), Porter is a strong proponent for the integrated study of human psychology and organizational behavior alongside core engineering fundamentals. In addition to technical strategy and software development, part of Porter's central focus at Amazon has been evangelizing how a deep understanding of one's emotional self as well as the emotional properties of others is a defining quality in successful technical leaders. Porter is a tenured member of the Executive Mentorship Program for the Foster School of Business at the University of Washington. He also serves on the advisory board for the Russ College of Electrical Engineering and Computer Science at Ohio University. Porter received his Bachelor of Arts in Computer Science from Earlham College.

Index

Printed in the United States
by Baker & Taylor Publisher Services